How to change . . .
DOG BEHAVIOUR

Part 1
ESSENTIAL THEORETICAL UNDERSTANDING

Peter Lewis

Illustrations:
Peter Lewis

A CANINE PUBLICATIONS MANUAL

CANINE PUBLICATIONS
21 Burridge Road, Burridge, Southampton, U.K. SO31 1BY
Tel: 01489 885112

HOW TO CHANGE DOG BEHAVIOUR
Part 1 - Essential Theoritical Understanding
Part 2 - Non-aggressive Problems
Part 3 - Aggression Problems

ISBN 0 906422 20 5

First published 2006

Printed and bound in Great Britain by Canine Publications
Typeset by Canine Publications

Dogs are far more straightforward than man for we are the devious species. They have taught me so much and therefore I dedicate this book to “Man’s Best Friend”.

Also available in the same series:-

HOW TO BECOME A DOG

HOW TO HAVE A PROBLEM FREE PUPPY

HOW TO TRAIN YOUR DOG - Part 1

HOW TO TRAIN YOUR DOG - Part 2

HOW TO MAKE GARDEN AGILITY OBSTACLES

HOW TO TEACH DOGS AGILITY OBSTACLES

HOW TO WIN AT AGILITY

HOW TO JUDGE AGILITY

HOW TO TRAIN A WORKING TRIAL DOG

Contents

ABOUT PETER LEWIS

Peter has a wealth of experience in training dogs and understanding their behaviour. In fact he has devoted 45 years of his life to the subject. He started training dogs for competition in the early sixties and quickly realised that there had to be better motivational methods than those largely used at that time. This search enabled him to develop many of his own revolutionary training methods as a very successful Obedience and Working Trial handler.

Then he applied his training talents to the then new sport of dog agility and was responsible for initiating and developing many of the motivational training methods still used today. He was thrilled when Agility commenced its own world championships, witnessing the culmination of many years pioneering work in that sport.

It was therefore a natural progression for him to help people with problem dogs and soon he was working totally professionally in the field of dog behaviour. Such was his success that over 50 veterinary surgeons were referring concerned owners to him.

Again he looked for better ways to understand dogs and to be able to rehabilitate them and thus greatly improve their behaviour. By applying his mind to improving techniques, he has therefore been able to help to extend the boundaries of dog behaviour knowledge and of solving consequent problems. Much of this he puts down to what he refers to as his solid apprenticeship as an enthusiastic dog trainer. During that time he judged and taught all three training disciplines across the world.

Still his fertile mind saw that Mary Ray's musical dog training demonstrations could become another competition discipline with the Kennel Club as its governing body.

Persuading Mary to join him in what was a new venture in 1996, the sport of Heelwork to Music is now well established in the UK and many other countries of the world.

Although the vast majority of his dog training years have been spent hands-on with dogs and dog people, he has also found time to write this, his twentieth book about dogs, which he feels has been the most important of all of his dog work. Whilst many people have written books about dog training and behaviour, few have the vast all-round experience and knowledge of the dog and how he acts and thinks than the author.

1

INTRODUCTION

I have been fortunate to write many books, but this trilogy on the subject of dog behaviour is for me the most important of them all. My reason for making such a statement is that it is the culmination of three factors which, once brought together, enhanced my ability to understand dogs.

The author

Those factors are in excess of forty years as a dog handler and instructor, studying the written word of dog behaviour, then finally a strong desire to find better ways of training. I had already put behind me 25 years as a practical dog trainer before I studied dog behaviour. To me those 25 years had been my apprenticeship where one learns what is practical and also develops their own theories of why dogs do certain things. I have no doubt that I shall continue to learn for there is still so much more that we do not know for sure.

I have always been innovative by nature and have put into practice many ideas. Not all were dog related and neither were they all successful, but if you do not try them out you will never know which deserve to succeed. Therefore when reading much of what had previously been written about canine behaviour I was able to use my experience to dissect the written word and think about it applied practically. Some I found to be excellent new knowledge, and some I found to be wonderful theory but either with missing parts or showing a complete lack of understanding the practicalities of dog handling. By thinking through problems my innovative nature came to the fore and I know that the results have helped some of our canine friends to be better understood by man.

When I decided to try and help pet dog owners to find solutions to what had already become, for them, a major problem I had to start at the ground floor. Despite any reputation I had in the competitive world of dog training I needed to convince veterinary surgeons that to place their clients dog's behaviour problem in my hands would bring about positive results. My name meant nothing to them and as my intention was to earn my living by working professionally at solving problems I needed their confidence. After all I had no degree in animal or dog behaviour but just vast handling experience backed up by my own studies of the subject and a desire to add to the fund of knowledge available to man about the behaviour of dogs.

I started dog handling in the early sixties and much of what was taught, almost without exception, relied on compulsive methods of training. Such compulsion was often clumsy and inappropriate for many dogs. Yet, for the vast majority of people, it was the only known method of dog training. There were odd exceptions and I instinctively knew the first two dog training clubs I attended left a lot to be desired. Luckily I found the man who was to set me upon a path of understanding dogs. His name was Charlie Wyant and he was a trainer of dogs for competition obedience. Way ahead of his time he thought much about better methods. With great difficulty he straightened my thinking and showed me how dogs could be taught using a combination of incentive, play and habit. He taught that in starting off puppies in this way most dogs could be trained to the highest competition standards without compulsion being necessary at all.

Armed with this information I set about trying to take such methods further and indeed I subsequently trained dogs to the highest level, sometimes hardly ever using a lead and collar. When dog agility competition started my outlets for finding better ways of training it were numerous, for it was a sport that cried out for motivational methods as the dog needed to work with great enthusiasm. My mind went into overtime devising techniques that would bring such results.

Eventually it came time for me to study dog behaviour to help solve problems with pet dogs. Again I looked for better ways of understanding and helping each problem dog. I say helping for sometimes owners can be driven to abandoning, re-homing or taking the dog to the vet's for his last journey. I reasoned that with better knowledge much of this could be avoided and thus began the most rewarding part of my life with dogs. I do not mean in a financial manner, but in putting a smile back on the face of so many owners while helping the dog at the same time.

I do not believe that dog training and dog behaviour can be separated for they are so interlinked. It is possible to study animal behaviour at degree level with no previous handling experience, particularly of dogs. It is therefore also possible, and often done, to set up a professional dog behavioural practice still without having acquired any handling skills. Needless to say such behaviourists will advise very few handling techniques as a remedy for they will be unable to demonstrate with requisite skill. It is all very well advising a course of action in an effort to provide a remedy for a problem. However, it is far better to be able to demonstrate the technique, thus ensuring and proving that the dog will respond.

To turn a person into a skilled dog handler, the first attribute they require is the latent ability to be able to understand what they are taught. Secondly they need to start instinctively using this aptitude while possibly forming their own ideas on modifying what they have been taught. Such a person, away from their instructor, will know what to do, when to do it, and how much of it to do because it will now be instinctive. There is only one way to put it and that is to say that person has got what it takes to be a successful dog handler at whatever type of dog handling they put their mind to. Here I am talking about people aspiring to greater levels of skill than that necessary for a pet dog owner.

Why is it that when we teach a hundred people with each one having the same information and having put in the same amount of work we cannot make a hundred quality dog handlers. More likely the figure will be less than ten and it would certainly never exceed 50%. Either a person has it within them to be an opera singer, an artist, a bricklayer, a gardener, or a dog trainer or they do not.

I am sure that everyone has something they are good at inside them but for many it will be a question of finding it. Therefore when a university or similar teaches animal, or in particular dog behaviour by lecture and the written word only, we will never know how many square pegs have been placed in round holes as a result. I am more than happy for people to undertake such studies and become behaviourists but they must also be instinctive dog trainers to be successful. If they are not then they will usually shy away from advising anything that might require a practical application and therefore demonstration for fear they will look foolish. Having said that, the boot should also be on the other foot for, as a successful dog trainer, it is a dereliction of duty not to study dog behaviour before practicing as a dog behaviourist.

Understanding every possible accepted proven theory and training the mind to analyse available information is another necessary attribute. To be able to genuinely help people whose dog is as important to them as their child requires practical handling ability, with sound theoretical knowledge and an understanding of canine psychology. If one link is missing the behaviourist is not complete.

So, as a veteran dog trainer and someone who has studied dog behaviour, I have set out to put on paper my experience. I do not claim all of it is my own thinking, far from it, but there are areas where I hope I have been able to improve understanding. While sometimes my methods rely solely on psychology, there are many that include a practical application. Neither have I completely thrown away discipline, for it is my firm belief that much of the troubles with western society is that young humans are allowed to express themselves, often to the detriment of others.

Mother nature made us how we are for a purpose and part of that purpose was to be competitive. Effectively that rules out the idea that all can be equal, and when we try and go against nature she always comes back at us, sometimes with a vengeance. So it is with dogs, for they were made to be competitive but with far lower levels of intelligence and without our ability to communicate. There will always be those humans and dogs that have more than others and there will always be canine discipline within a pack of dogs. We would do better to acknowledge this rather than to try and know better than mother nature.

I have not set out to write a scientific thesis and I have written this book in the manner in which I speak with my clients, which is avoiding all but minimal behavioural jargon. In this way I hope that the reader will more readily understand what I am trying to convey. I do ask for one indulgence to be tolerated and that is I am a great user of scales as a method of explanation. Therefore when the reader comes across yet another scale please bear with me. My intentions are that this book will assist those studying with a view to practicing as a canine behaviourist, and also to help dog owners to avoid their dogs developing problems.

Understanding dogs is the first secret and indeed another book in this "How to" series is entitled "How to Become a Dog" which aimed to help owners stop thinking like a human and use canine logic instead. I have already stated that some of what I have written is not original thought for there have been others that have influenced me. Previously I have mentioned my greatest mentor, Charlie Wyant, but others whose books made me think a lot were those of John Holmes and John Rogerson. On the subject of dog behaviour, John Rogerson is correctly recognised as a world expert. He is another example of a practical dog trainer who has studied and thought about his subject and I commend his books to you.

Sometimes my clients want to reproach themselves for getting things wrong with their dog and I always tell them they cannot expect to be an expert at everything in life. Nor can I, for there are many occupations I would have been very inadequate at. Therefore to admit to having a problem with a dog is not an admission of failure but rather a cry for help and knowledge from someone who may have the skills required.

It must be remembered that this trilogy is not meant to be a basic dog training book. I and others have written many of them. This is about modifying dog behaviour when all else fails and therefore avoiding such problems in the first place.

2

KNOWLEDGE AND LEARNING

It is fairly well documented that the domestic dog is a descendant of the wolf. As much has been written on this subject I will not do the same. Suffice it to say that a lot of what applies to a dog would apply to his ancestor, except that all those years of living closely to man has changed some of a dog's characteristics. However, this first part is about essential theoretical knowledge and understanding the dog. Theory may be boring to some but it is the basis of understanding.

A DOG'S MIND AND THOUGHTS

Compared to what we can think about, a dog's thought processes are extremely limited. I have already suggested that we learn to think like a dog, so what will that mean to us? Suddenly we will find that the secular world we live in has little use for us and our values are very different. No longer are we concerned to have the latest mechanical or electrical gadgets, or about the style or quality of our clothing. Possible wear and tear on our home furnishings is of no interest, in fact we might indulge in hastening such human problems by the use of our teeth.

Making up our days of thought will be all the survival issues such as food, water, rest, sleep comfort from the elements, even though we may be a family dog. Of course survival issues will have offshoots, such as a desire to hunt, and this will vary greatly in its intensity from one of us to another. Then there will be issues of fear or concern such as feeling personally threatened or possible threats to the pack's territory. For some of us, being seen to be important in the pack structure will loom large in our thoughts.

At certain times procreation may well come to mind, and very often possible interaction with other pack members will be issues our brain must make decisions on. It might be their close proximity, affection or use as a possible playmate. If during each day the subjects we think about just exceed the number of fingers on a human's hand then we have really had a big day. On the next day it will be most of the same things thought about all over again.

Thinking

Human thoughts include all the aforementioned but, as man progresses, his thoughts include a host of different sophisticated ideas which have far greater offshoots.

Perhaps the greatest difference between the two species is that one can explain and reason and the other cannot. Therefore when mother says to little Johnny "Now come on, be a good boy" because of past explanations it means to him stop misbehaving. To Fido who has understood by habit what "Good boy" means, his association of the simple phrase is carry on misbehaving.

That is a simple explanation of the difference, but one can expand that to so many factors that affect our relationship with dogs. Their lack of inability to reason and work back through a sequence of events and thus understand the consequence of them is borne out in several ways.

A classic is the inability to comprehend why we might be consistently angry every time he leaves a noxious deposit on the floor of our house. All that he understands is that the deposit is there and we appear to be angry.

He cannot think through his behaviour that led to him leaving it there. If he could, then teaching dogs to be clean in the house would be quite simple. All we would need to do is to show him the deposit and speak angrily. Many people do just that but wonder why the dog does not get the message.

HOW A DOG LEARNS

A dog learns in various ways. He can learn from his senses for they will trigger instincts. It has already been established that instincts are inherited but that does not stop the dog learning as a result of using them. His predatory instinct may lead him to chase small animals but he will learn that while he may be able to catch a rabbit he is less likely to catch a squirrel. That does not mean he will eventually stop chasing squirrels, for much of the enjoyment is the thrill of the chase and making the animal back down by running away.

He can learn by watching other canines, but when no human or canine influences are available and he tries something never previously attempted he may make several abortive attempts before finding the easiest solution. That might be by luck, and an example of this is when left alone in a room and he repeatedly jumps up at the door to reunite himself with his pack. Immediately, and definitely inadvertently, his paw catches the door handle which has the effect of opening it.

Hunting instinct

Dependant upon his level of intelligence, one or more future attempts will finish with him aiming for the handle with the paw rather than jumping up.

As a puppy he will learn much from his mother, brothers and sisters. In the nest games will be played which will teach many things. He will have his first lessons in canine social interaction and ranking order. His investigative instinct will teach him many things. Some will give him pleasure while others will have an opposite effect, and therefore it is easy to understand which of those two he is likely to repeat.

Because we love the company of dogs so much we domesticate them and then find that we need to train them, so in effect we become their tutor to teach them things, many of which they would not need to learn without us. Dependant upon their intelligence levels allied to other factors of their make up, some will learn more quickly than others.

There is no doubt that there are great differences with learning abilities between different breeds. However one of the factors that affect this is the rank of the dog. This is discussed in depth starting on page 38.

What I mean is that while a high ranking dog may be extremely intelligent his perception of pack superiority may cloud the issue for it is more important to him not to obey an inferior. However, suffice it to say that, generally speaking, dogs that have a history of being worked by man generally have a greater desire to learn when we try and teach them. This generally means that such breeds are more receptive to training. However by reading on you will realise it is not quite as simple as that.

Many years ago there was a study carried out by Oxford University on the most intelligence species known to man. Their conclusion at the time was that it was a Border Collie. One can only assume they realised that different breeds had differing intelligence levels for they did not give dog as the answer.

Having owned several Border Collies I have to say I agree with their findings, for there is no doubt they are the fastest learning breed of all. Do not rush to get one for they are better in skilled hands that understand they can learn all the wrong things very quickly. In other words unless you really understand dog training owning a Border Collie can become at the best a frustrated dog and at the worse a liability.

GOOD OR BAD RESULTS

The simplicity of a dog's mind I liken to that of a baby of less than a year of age. In other words explanation of what we require or do not require is impossible and, therefore, they tend to work on the simple premise of good or bad result. If they carry out an action and it gives them pleasure then they are certainly likely to repeat it given the opportunity to do so. On the other hand if their action brings about an unpleasant result they are less likely to want to repeat it. For that reason earlier dog training was based upon correction and reward. Usually reward was just praise, and correction was being checked by the lead. This meant the trainer could apply a good or bad result to the dog's actions.

While methods of dog training have moved on from that simplistic form, it is still valid when all other methods are unsuitable. It can also be skilfully used in an entirely non physical manner. For instance, if the dog seeks attention and doesn't get it, that is a bad result.

If the dog is consistently ignored he will eventually give up seeking attention. Of course eventually granting the attention he previously sought but has given up on will bring about a good result, and the dog will have started to learn that attention to him more often occurs if it is on the owner's terms.

HABITUAL LEARNING

One of the greatest ways a dog learns to repeat behaviour is by habit. In fact that is how we train dogs. If we consistently get the dog to repeat a behaviour we want him to learn, then by adding command, signal, general body language or all three to that action, he will eventually do it upon command or signal, etc. Additionally, if we show him what we want him to do and consistently add a trigger such as set up, body language, signal, command, some or all of them, then by habit he will also learn our requirement.

If we include motivation, learning becomes pleasant and more quickly absorbed. Likewise, if the dog chases a squirrel for the first time and gets a good result from his behaviour, be sure that the next time he sees one he will repeat that behaviour more readily. At this point, part of his behaviour is driven by his predatory instinct and the other part is driven by what is starting to become a habit. If the dog climbs on the furniture for the first time and gets away with it, expect habitual learning to be starting to set in.

When with a client and I am dealing with their dog's inappropriate behaviour then I need to ascertain the cause of that problem. Knowing what the cause is will make it far more likely that I can advise and/or demonstrate the best possible way to change the behaviour. However, I have to recognise that I am also dealing with habit. This is the dog's simplistic mind telling him that "under these circumstances this is what I do". Once again it may be a straightforward case of see squirrel, chase squirrel. Therefore I have to ensure that the cure I advise deals with both the cause and the habit.

ASSOCIATION OF IDEAS

A dog also learns from an association of ideas born out of habit. This might be sequential learning several events which are associated together in the order in which they are taught. This is where the dog will know by habit what comes next in the sequence. I am tempted to call it chaining but that is dipping into behaviour jargon. However, in this instance its use is quite descriptive for we can visualise the links making up the whole chain of sequences.

As I write I have a 20 month old Labrador bitch called Sassy who knows exactly the sequence of events in the morning because, like many of us, both my wife and I are habitual ourselves. I go to the toilet and the dog hears the flushing, etc. I let her out of the kitchen and take her out to be clean. Note: take her out not send her out. I leave her in the kitchen again while I get washed, etc. Then I make my breakfast and when it is ready start eating, having given her two small sausage roll biscuits before I do so. I finish breakfast then she has almost 50% of her daily feed. Following this she has a bowl of water with a dash of milk. Next she accompanies me to the bedroom to help me get dressed. Finally I take her out for a walk.

All of those parts of a sequence of events are exciting to her and by choice she would anticipate each one but she has to wait for my timing. The whole sequence is therefore a chain. If on odd occasions I do things out of sequence she will clearly display mild confusion by trying to guess what I am going to do next. This might be rushing in one of the directions that are normally part of that chain.

Looking at kitchen door

We can therefore train our dogs by association of ideas, and a good example of this is the sport of dog agility whereby the dog always knows there is another obstacle to be jumped until there are no more available obstacles.

Most modern dog trainers will tell you that if you send your dog into the garden to relieve itself, then when it returns and you give it a treat you are not rewarding the act of being clean in the garden. Rather you are rewarding the dog for returning as that was the last thing the dog did.

In fact I have followed this principle in the past but I would now like to debunk such theory on the basis that the act of being clean is but a link in the chain that will culminate in a reward. Therefore, it is valid providing that the owner always knows that the dog has been clean, otherwise a greedy dog will soon learn to cut out that link in the chain and return just for the food instead of getting on with his duties.

CIRCUMSTANTIAL LEARNING

Part of this association of ideas is connected to the set up and it can be called circumstantial learning. If I take my dog to another house then she will not so easily expect the next part of the sequence to occur. That is because, inadvertently, I have only taught her the sequence in my house. Therefore that tells us that we have to teach the dog in various places and venues to understand the requirement applies anywhere that is similar but not necessarily the same. If I only ever teach my dog to climb a six foot scale jump in my garden then I cannot expect her to understand the requirement until I have taught her the similarity at a variety of venues.

When the pet dog owner has reasonable recall control of their puppy in the house and garden they expect that will be the case the first time they take it to the park and let it off lead. This is where most come unstuck straight away, for puppies of average insecurity levels or less genuinely do not know what to do. Although they know their name and maybe a recall command used at home, it means nothing in the park until taught at that place, and preferably at several other venues as well. Those puppies of higher insecurity levels will recall more easily, but that has little to do with teaching and more to do instinct to stay with the pack. For more detailed information on this subject see Part 2, Chapter 6. Therefore be careful not to think you have successfully taught a recall in the park when in fact nature may have done it for you.

Set-up can also allow us to use the same command to mean different things to the dog. It is similar to association of ideas in that it relies on learning a set of circumstances, hence my terminology 'circumstantial learning'. It is also where everything has to come together to mean a certain thing, but change one or more parts of the circumstances and the dog can be taught that it means something different. An example might be the sheepdog trainer who commands "Get down" to the dog while he works the sheep. However, the dog doesn't actually go down but rather crouches more, therefore slowing down his current busy way of working. That will be what the shepherd requires. Having finished working the sheep and the dog is with him and he tells him to "Get down" then he goes completely down. In other words the command is the same but the circumstances are different and they mean two different things to the dog.

Mary Ray is a great competition dog trainer and will use the command "Back" to mean two entirely different things. One, during competition heel work, will mean "do a precise right angle left turn with your body without in any way cutting the corner" and the other will mean "turn to the left in competition agility". In the first case the dog is close at heel and her hand will be placed on the left side of her body around the front of the waist. That is her set up for competition obedience heel work. While running her dog in agility her hand will not be in that place and the whole of her body posture will be different. Therefore the same command means different things to the same dog.

I can also put that example round the other way. When I say "Steady" to Sassy she will walk loosely but not accurately on my left hand side. I use that command on a walk when I feel it to be necessary for her to be walking reasonably close to me without great accuracy. Usually it is for low level safety.

When I want her to undertake competition heel work I have a different set up, for first I will tell her to "Watch me". The degree to which she does this is not important but it is part of the association of ideas. Then I tell her "Heel". In both cases she is walking at heel but one set up means accurately while the other means loosely, although in this case they have different commands. Also my body language or posture differs from one command to the other which is added information for the dog.

TRIGGERS

It could be said that commands and body language are a type of trigger which means something to the dog. A trigger is really another word for stimulus for it describes a response, to certain circumstances. If the lead is only picked up when taking the excitable dog out for a walk then expect him to be excited and almost unmanageable from the time you pick the lead up. If when the door bell rings the dog instantly barks, then expect him to repeat that when the same ringing sound occurs on the TV or Radio.

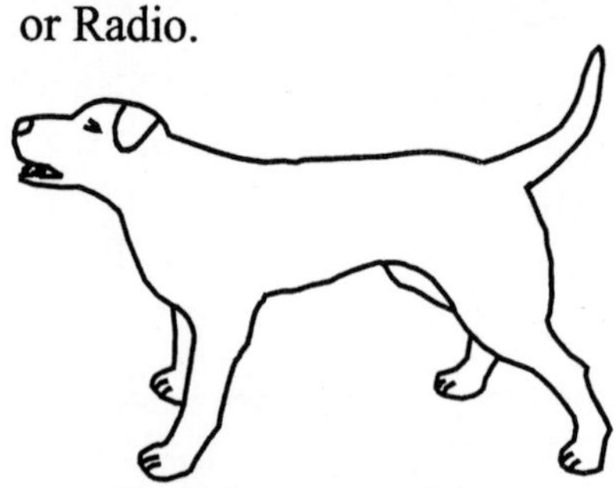

Barking at callers

There are also visual triggers and, of course, triggers can be smells and literally anything connected to the senses, instinct, or what they have been taught. They are always something which reminds the dog of a previously learned set of circumstances and his past response to them. Therefore if you want him to change his behaviour to a trigger, use it in circumstances where what the dog expects to occur does not occur. On the other hand if you want him to learn what a trigger means, then consistently apply it at the appropriate time.

Recapitulation

All of these facets of learning can come together to teach the dog and, to remind the reader of what we have dealt with so far, they are:-

Good or bad results.

Habitual learning.

Association of ideas.

Circumstantial learning.

Triggers.

NEVER LET MISTAKES OCCUR

Bearing in mind all the foregoing, it is easy to understand that if a dog never makes a mistake he cannot learn the wrong thing. In other words we have every opportunity to ensure that the only things he learns are those that we want him to. I have to admit that it is an impossibility to carry out such an aim and therefore gain 100% success. However, it is like anything in life; if you only strive for mediocrity then that will be the maximum you attain, and it may well be less than that. If, on the other hand, you make as a target the highest possible attainment then you may get close to it.

Such high ideals always mean much work and a lot of inconvenience but the results can be really worthwhile. In such a way I have trained quite a few of my own dogs and generally I have been able to attain a figure of success in the low nineties.

The odd mistakes that have occurred have not been regularly reinforced, so therefore mistakes that have been made have not been regularly repeated and success has been the result. By success I mean dogs that have been trained to high levels of competition without other than minimal vocal correction being necessary on odd occasions.

Of course I must qualify this to say that I would believe that only those with much handling skill would be able to reach this ideal, yet it is still worth the pet owner trying for it. Maybe success might be reduced to 70% but that is still a better figure than 40 % or less. This is a big reason why those who understand dog behaviour and have vast levels of handling experience are best qualified to deal with behavioural problems. Their perception of when the next step leads to a mistake will have been honed by years of experience and this cannot be learned by study. It becomes like an uncanny sixth sense which, when allied to the knowledge of canine behaviour, is a powerful experience that underpins the understanding of dogs.

3

SENSES, NATURE AND INSTINCTS

In the previous chapter I have made a few references to instincts and senses, so now is the time to list them with brief explanations.

SENSES

To train dogs we need to understand their senses. The most important of them are:-
Smell - Sight - Hearing - Touch - Taste. While we have these senses, the ability of them differs and they are used in different ways. For a start our intelligence and power of reasoning allows us to evaluate what our senses tell us in very sophisticated ways. With the dog they are used as a much more basic tool.

Smell

Smell is the sense that the dog relies on most. It is far more developed than a human's, in fact it is difficult for us to comprehend just how acute this sense is. Dogs use it for a variety of information. One is the amalgamation with the investigative instinct which can lead to satisfying the eating and drinking instinct. Sense of smell can also detect possible danger and is often used to identify friendly beings. Also there is man's use of the dog's nose for particular purposes

Sight

Dogs see in a different way to humans and while they easily spot movement they are very poor at identifying detail. Some dogs have very poor eyesight but don't show it as much as we would for they compensate with the heightened senses of their nose. It is thought that the dog sees things in shades of black, grey and white.

Hearing

Sound sensitivity is a subject dealt with in this chapter. Suffice it to say that a dog's hearing distance is similar to ours and, where there is a difference, dogs can hear from a greater distance than we can. Certainly they can hear much higher pitched sounds than we can.

Because we can understand what many acute sounds mean we are less frightened of them than many dogs that often only have a comprehension of danger triggered by particular sounds.

Touch

Touch has more relevance to a dog when he is touched by others. Such touch will mean different things to him dependant on his mode at the time. For instance, if he is in a fight and he is touched from behind then he will likely attempt to bite whatever touched him. However, if he is in gentle mode then he will expect our touch to mean reassurance.

When he touches, it is usually without a lot of finesse. However, a dog may attempt to touch another with his paw in quite a gentle manner. Touch is not only used with the leg but also the mouth, nose and sometimes the sides of his body.

Taste

Using this sense certainly allows differentiation between flavours, and dogs will often display a liking for certain tastes. This can lead to the unwary owner having to feed the dog with best steak as he has learned that to refuse other food will get him the taste he prefers.

NATURE

For many years I have thought about a variety of things connected to this world that we live in. Some time ago I came to the conclusion that nature is much more of an overriding factor than we often give her credit for. Think of all the things where man has interfered with nature. So often she comes back at us and sometimes with a vengeance.

Take the complications of the British climate, arguably one of the more difficult countries in the world where weather forecasters are concerned. Yet if we have a long dry spell it is almost certain that some time in the not too distant future we will have a long wet spell. For me that is nature carrying out her balancing act.

There are so many instances other than climatic changes where nature plays her part. I ask myself why does she introduce new illnesses that at the time we have no cure for. Were we meant to invent penicillin or did she decide we need super bugs that are not resistant to that antidote. Therefore I firmly believe that this rule applies whenever we try to alter what she wants to happen and this happens with the breeding of dogs. Why did hereditary defects come into existence, and why when a breeder tries to determine the likely character of a litter of puppies does one or more in the litter have differences.

There are many characteristics that make up the nature of a dog. Bold or weak - pushy or retiring - worse still dominant or wimpy. Almost secure or insecure. Fearful or bold. Excitable or laid back. Easily distracted or not. Active or sedentary. Loving or independent. We need to identify all of these parts of a character for nature may well have made the decision on behalf of the breeder. Of course many matings take place without a breeder deciding who shall be the parents but still someone has to decide the characteristics of the offspring. Therefore at times in this book I will refer to the interference of nature

INSTINCTS

Much of what a dog thinks about is driven by the instincts given to him. For our purpose they can be further described as inherited behaviour. In some breeds various instincts have been enhanced as a result of hundreds of years selective breeding. Such inherited behaviour can be of great advantage to man and others. Enhanced problem levels of it can equally be, at the least, an annoyance for those of us who would rather the dog did not have it. Understanding canine instincts is one of the pre-requisites to understanding dogs and their motives, and for those wishing to train dogs a deep knowledge of these canine traits becomes essential.

Intensity of instinct

There is also nature's intervention when deciding that one or more of a litter will have enhanced levels of one or more instincts. Take for instance the predatory instinct which starts with the thrill of the chase and may end with aggression. If enhanced above an average level it can result in the dog wanting to chase virtually anything that moves quickly. Therefore I like to add to instincts a scale describing levels of intensity. One to ten is ideal and obviously five becomes an average score. While the dog that chases everything will score at least nine, those that display little interest in chasing will score three or less.

Pack instinct

This is the instinct to herd together which is far greater than that given to our species. Dogs are therefore social animals who want to be part of a group and some develop friendships. This instinct is also there to help protect against predators, for successful defence when attacked is greater with numerical strength.

When there is more than one dog together they become a pack. It is then that canine social matters dictate by instinct that there should be an order of importance within the pack. Such an order can evolve by either aggressive or non aggressive methods. Which of these two prevails will depend upon the natural born rank of each dog. The higher the rank the higher the likelihood of using aggressive methods to impose will, while lower ranked animals will only fight for survival when threatened. This subject is looked at in greater depth in chapter 4.

A relaxed pack

Territorial instinct

Dogs are inherently territorial in that they instinctively define an area which they believe to be the territory of their pack. Outsiders to the pack are not welcome and will be made aware of this by one or more pack members barking at them and possibly showing higher degrees of aggressive behaviour. Regular non aggressive visitors to the pack can be welcomed but it will depend upon the frequency of their visitations and demeanour as to when they are regarded as trusted.

When the level of intensity of this instinct is enhanced the greater becomes the display of territorial behaviour. When it is well below average the dog can show little or no territorial behaviour. The second and third parts of this trilogy look at the problem and possible curtailment of territorial aggression.

Guarding instinct

This instinct can often be allied to that which makes the dog want to guard his or the pack's territory. However, the dog can guard other than territory. For instance he can guard a pack member or members away from their own territory and also he can guard a possession that he has claimed for himself. Guarding is therefore usually associated with aggression, for if the ultimate sanction is not to be used there is little point in guarding in the first place.

Barking instinct

Barking is often as a result of territorial and/or guarding instinct being triggered, but not always for there are other reasons why a dog barks. Dogs will often bark for attention; as a call to others, as a warning. Then there is excitement and when there is a perception of being threatened. They all have different sound structures and can be identified by careful listening. There is also barking associated with insecurity while separated from the pack, but not all dogs are affected in this way.

Sex instinct

Both humans and canines are given this instinct as a method of ensuring that the species continues. It is there for procreation and while the male will always take the opportunity to mate, the female only does so at certain times of the year and roughly for a few days in the middle of her season. At this stage she becomes like the male and will virtually mate with the first one to come along. This instinct can be over average and therefore cause us unwanted problems.

Cocking the leg

When we have male puppies we wait almost with bated breath for the first time that the puppy ceases to squat for urination but lifts his leg instead. It is an instinct that is not realised by the dog until his testosterone levels are becoming mature. Usually the male will lift his leg against something tall and a classic example is a tree or bush, and this can only be his instinct to leave a message for all other canines or animals in the area.

Predatory instinct

This is a basic instinct for, in the wild, hunting for food is a pre-requisite to survival. When the dog spots an animal, particularly those moving, it will initially stare then possibly stalk it carefully and/or immediately give chase. If he is lucky enough to catch it there are several possibilities. He might bite it hard enough to kill it or just hold it. Then again he might shake it or just play with it. This will depend upon two factors which are the breed and therefore nature of the dog and the degree of intensity of this instinct. In most cases he will lose interest or the prey will escape before he catches it, but each time he indulges in this instinct he reinforces it and thus it becomes a habit.

All the retrieving breeds will usually have sufficient levels of this instinct to indulge in the thrill of the chase, and this will also apply to the herding breeds. These are the groups that are less likely to carry on through to the killing stage.

On the other hand terriers and hounds are far more likely to go to the last stage for this has been a requirement of man through hundreds of years breeding. The groups of breeds mentioned is not an exhaustive list, as any breed can have over average levels of predatory instinct which will lead to this type of behaviour problems. Predatory aggression problems and solutions are dealt with in Chapter 5, Part 3.

Killing instinct

This instinct might be present in all dogs but its level of intensity is usually well below average. We have already discussed how the desire to kill as the climax of a predatory chase is the exception rather than the rule, but it is often the only time that killing will occur.

More hunting

Of course there are instances of dogs killing for other reasons and even some of the few incidents might be the result of accidental killing. This can be the result of inadvertently gripping a being too hard. However, when an animal indulges in this basic of instincts, man rarely attributes its occurrence to an accident for while a dog is a dog, a dog that kills is a killer.

Fear Instinct

The vast majority of dogs have this instinct. Some dogs will not display fear for many years of their life and then something happens that will trigger it. Usually from then on it also becomes a habit. Therefore my own feeling is that the dog is born with minimal or far greater fear levels even if not yet triggered. It has been established that three factors will most likely bring out fear in the dog. The most obvious is the passing on of fear through the genes of one or more of the dog's parents. Then there is the possibility that the mother has, inadvertently or not, transmitted fear of something to her puppies in the nest. There is also fear learned by a frightening experience.

Levels of intensity will be average, high or low, dependant upon what nature and breeding, has decided and also where there was a frightening experience how bad it was. Extreme fear will lead to flight with the most likely inducement being extreme sound such as bangs, thunder and fireworks.

Fear

Fear can also lead to aggression as a defensive measure and is arguably one of the major causes of dog biting incidents. Far greater information on this subject can be found in parts two and three of this trilogy.

Leading instinct

This is the ability to lead another pack member, canine or human to where something has happened that he knows will concern other pack members. It has been well documented over the years, often with the herding breeds alerting their shepherd to a lamb in trouble. While it is not a taught behaviour the leading instinct may well be greater in the dog who respects others in the pack to have higher status. In this respect the instinct is developed as an offshoot of learning.

Eating & drinking instinct

All creatures have an instinct to eat and drink for it is a basic necessity of life. Without water neither dogs nor humans can survive for long and food is necessary for wellbeing and energy. Once again this instinct is prone to levels of intensity, particularly where food is concerned. There are dogs who are naturally extremely greedy and often scavengers as well. The group scoring around an average intensity mark would be in the majority and of course there are also dogs at the lower end of the scale which only eat as a necessity to survival.

Sleep

Rest is also a prerequisite for life and is the most effective way to recharge the batteries. Dogs sleep at different levels of intensity, and this is a basic requirement, for in the wild to go totally to sleep could make the dog unaware of those that wish to do him harm.

Very often a dog will sleep in a strategic position just in case he does go into a deep sleep. This is best reflected in domestic situations whereby he will lie touching a person or the chair a person is sitting on in the knowledge that if they move he will be woken up. As an alternative, lying in positions where they will have to be disturbed should other members of the pack (family) move past is an often used option.

We must remember that we can ask other human family members where they are going. Dog's cannot do that so they need to watch us to ascertain what our body language will tell them about our intentions.

Dogs also dream and so often we will see the legs twitching during a dream whereby the dog is obviously dreaming about something which requires movement. Like us they can also make funny noises during their sleep and sometimes these will be noises that indicate worry or concern.

While puppies spend much time sleeping so do dogs as they become older, in fact many such domestic dogs seem to finish up just eating and sleeping but that can also be the result of us not giving them a more active life.

Investigative instinct

From three weeks onwards puppies display their investigative instinct and this is they way in which they find out much about the world they are to live in. One has only to watch the behaviour of a dog on a walk in the country to see that much of it is investigation. Sniffing will be a major part of this instinct albeit close to the ground and sometimes higher for airborne scents. They will be particularly interested in the urinary and anal deposits of

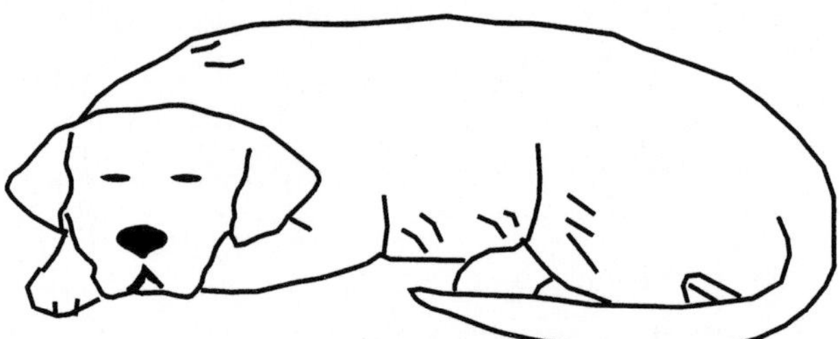

Sleeping

other dogs, trying to gain from the smells as much information as possible. They will also use their ears and eyes for investigation, with the hearing aspect often being quite visible when there is a sound different from normal. When they meet another dog there will be investigation of anal and genital regions of the other's body, while the same behaviour will most likely but not definitely be inflicted upon them.

Play instinct

The instinct to play is in all dogs but unless it is developed those with low levels of this instinct will probably finish up as adults with no interest in play whatsoever. Thankfully such dogs are in the minority, and certain breeds are renowned for being playful because of man's ongoing selective breeding. All the retrieving and herding breeds easily fit into this category. This instinct can be honed and developed and, as long as it is carried out during the puppy's formative months, the result will be a dog that is always ready for a game. For man's purpose of training dogs the development of this instinct is important.

Circling down

All dog owners are aware that many dogs will circle before lying down. Theories are advanced for this such as trampling down the grass. I would have thought that it is just an act of trying to make the ground or the nest more comfortable. However, if this is the case we must ask why they often appear to make the bed we provide less comfortable by indulging in this instinct.

Playing

4

UNDERSTANDING RANK

NATURAL BORN RANK

One of the major influences in a dog's life is the rank given to him by nature. For dog trainers I believe this to be the most important part of understanding dogs. Rank is a measure of how pushy or bossy a dog might be or indeed the opposite, such as dog who is always willing to allow others to be more important. Then there are all those in between these two extremes. In this respect we humans are not unlike as we are also born with differing levels of rank.

Disguised human rank

We often inadvertently disguise rank, such as when born to a powerful or wealthy family. Advancement in business or the services can distort the picture where we can be promoted beyond our ability to give the leadership associated with the rank conveyed upon us by others. Then there is our education and understanding of manners which, to a certain extent, teach us to try and suppress our desire to be better than the next man. With a dog such suppression never even enters the thought process and any dog quickly understands that another has inferior or superior rank.

DESCRIBING RANK

Usually when dealing with canine behavioural problems, ascertaining the dog's likely natural born rank is very important. In fact in some cases it can be crucial. While starting to study dog behaviour I realised that in the first place the attention paid to this part of a dog's character was minimal and that it would be easier described by using a table. After some thought one to one hundred seemed logical and the table which follows is the one I have used for many years. It is of great assistance to me when describing to my clients the rank of their dog and what this means to them.

Wimps to dominance

As an explanation, with a score of one we have the world's biggest wimp while at one hundred is the most dominant dog ever likely to be encountered. I have to say that I have never given any dog these extreme scores, with the highest score going to a very large male Greyhound rescued after being retired from racing. Three veterinary surgeons had refused to see him. None could get close enough to apply a muzzle. I was eventually recommended by a fourth veterinary surgeon and to this dog I applied my highest ever score of ninety-five.

I dread to think what a higher scoring dog will be like. But if ever it is my misfortune to meet such a dog I know that if I avoid interacting with him it is highly unlikely that I will be bitten. The trouble with this one was that I lost my concentration and made the mistake of attempting to take my report sheet, that he had stolen, out of his mouth. It was just a reflex reaction while continuing to talk to his owner. Result a severe bite on the hand for I had broken a golden rule with extremely dominant dogs, which is woe betide anyone who tries to take away from him something he has decided is his.

I was disappointed on two counts. One that I had lost my concentration, and the other was that it was the first time I had ever been bitten. I am pleased to say one bite remains the score to this day for I do not believe people working with aggressive dogs should bear many scars of their work. Quite the opposite should be the case, brought about by using their experience to know when the next step may well mean being bitten.

TABLE OF RANKING VALUES

This is an approximate guide that will be applicable to many dogs. However, other factors can influence a dog's score as seldom is anything to do with behaviour black and white for there are many grey areas that can confuse even the most experienced. Therefore all of a dog's characteer must be studied.

Value 1 to 19

Extremely low ranking

These dogs will usually also have high fear factors but that does not have to be the case. They are submissive to the extreme and cannot take any pressure. Requiring very gentle and supportive handling, harsh words and any discipline are entirely inappropriate. They can be allowed many privileges as they will not seek to take advantage. They are followers which have no desire for rank order advancement, always preferring to please.

Value 20 to 39

Low ranking

Only towards the top of this group will there be any desire to improve pack status and it will be minimal. Extreme weakness will need to be shown for these dogs to take advantage. They also often display fear but not as extreme as the lower group. They are very submissive dogs which cannot take pressure and, when being trained, require extremely supportive and careful handling.

Value 40 to 59

Average ranking

This is where the majority of dogs score and generally they are responsive to training and happy to comply. Towards the lower end of this group their attributes will be closer to the low ranking dogs. At the higher end they can start to seek advantage and a little more authority might be required to nullify any perception of superiority.

Value 60 to 69

Weakness aware

This group will always spot pack weakness and, where possible, take advantage. However, if this is understood by the trainer and superior ranking is displayed, they are easily controlled. Usually bolder than the previous group, they will still display submissive signs if the other is too authorative.

Value 70 to 79
High Ranking

In this group the higher the score the more difficult will be the dog to bend to your will. This group's type will take over if allowed to, particularly at the top end. They will often need much firmer control and quickly spot weakness. Unless experienced, care is required if their perceived right to do or not do something is challenged. This group are often erroneously described as dominant for they will still back down to a perceived superior. For me, dogs I class as high ranking usually take on some of the attributes of dominant dogs but to lesser degrees of intensity.

Value 80 to 100
Dominant

These are the most difficult of all to control. Their one desire is to lead the pack and do so unless one or more members of the pack are perceived to be more dominant and thus in control. The higher the score the greater the problem. When it exceeds 84-85 it usually means that we cannot possibly win a confrontation and should neither seek nor accept one of the dog's making. Only psychology can be used to make living with them acceptable, and at the very top end that may be impossible.

RANKING COMPLICATIONS

Dominant yet fearful

While fear is mentioned in the lower groups it is possible for a dominant dog to also display fear. So once again bear in mind that this chart is only a guide and other conflicting parts of a dog's character can lead to an incorrect grading. Rank can also change in different areas as I will subsequently discuss. Also through life their scores can change so that they move to a different group. This most usually occurs when a previously high scoring dog becomes physically old together with diminishing senses.

There are differences of opinion among those working in the field of dog behaviour. Some will argue that it is possible to change the dog's rank while others have the view that the dog is what he is and only his perception of our rank can be changed. It is this latter view that I mostly subscribe to but not completely for reasons that will become apparent a few paragraphs on.

A deceitful species

Homo sapiens are the most deceitful species even when we use this ability without malice. We call it acting. Therefore we can fool a dog into believing that our rank is far superior to his although he ranks higher than we do. It may mean acting as though in a temper, but to be effective it must be carried out in cold blood.

Actor

This is because following such a display it may be immediately necessary to be supportive to the dog. If having the real emotion of temper that becomes impossible.

Acting is not something dogs do but there can be confusion for us when two or more interact and we must decide which the higher ranked dog is. One may be pushy while the other seemingly allows it to occur. However it may be the latter one that eventually puts his foot down so it is really a question of observing closely over a period of time to find the higher ranker. Remember that it will be the question of which one finally backs down to the other that determines superior rank.

Grey areas again

As usual grey areas can come into play as rank can appear to change from one place to another. It was established a long time ago that rank can change territorially to non-territorially, and while I have observed many instances of this

I have also seen some instances of it changing from one room to another.

I will give examples. I had two dogs at the time, a bitch scoring 79 and a wimpy male scoring 35. They were the territorial scores. Upon passing through a hedge row at the end of my land the positions of ranking superiority reversed, because off territory the fear of the unknown affected the bitch far more than the dog. He in fact appeared to be bolder while she became wimpy and followed his lead. This situation reverted to normal as soon as we crossed back through the hedgerow on to our territory even though we may have been on a walk for two hours.

Wary in the lounge

I currently have the same, but now older, low ranking male and a different higher ranking bitch with a score in the high sixties. She bosses him around whenever she feels it necessary, and that is several times a day. There is no biting but she will give him a hard time. However, in a certain part of the lounge of the house she is very wary of him. This is easy to understand for as a puppy, in an effort to retrieve a thrown toy; she accidentally jumped on him while he was resting. He displayed much instinctive displeasure, chasing her away while growling. This happened on two occasions. Now she is wary of trying to retrieve a toy close to him when he is lying approximately in the same place. If only that low ranking dog would sometimes take the same action with her in other parts of the territory he would have less superior displays from her. Of course she continues to act out of habit as past circumstantial learning brought about a bad result in that place when the same set up occurs.

Gender affects rank

I have brought gender into the equation and that also has an effect upon rank. Generally speaking, and with the possible exception of the more dominant dogs, males will back down to females for some of the time but not all of the time. This is not unlike human relationship. Therefore a male scoring 40 and a female scoring 60 will, at times, appear to have a bigger gap between their scores. Reverse the genders with the male having the higher score and sometimes they will appear to be more closely ranked.

There are those dog owners who in the confines of the house are able to get across sufficient levels of superior rank but who rarely interact with the dog in the garden. Therefore the dog will believe that in the garden he has no need to follow instructions given to him, and that his two legged pack superior in the house has far less power to impose will when he is running free in the garden. Needless to say the larger the garden the bigger the problem.

Recalls in the park

Why do so many people have such trouble recalling their dog in the park and similar wide open spaces while at home they usually get a good result. It is purely because the first time the dog runs free on a walk he is allowed to realise that the power he knows his owner possesses in the house diminishes rapidly when there are no confines. There is also the habit factor and circumstantial learning to take into account, for we should not assume that as he returns when called in the house he will do so in the park.

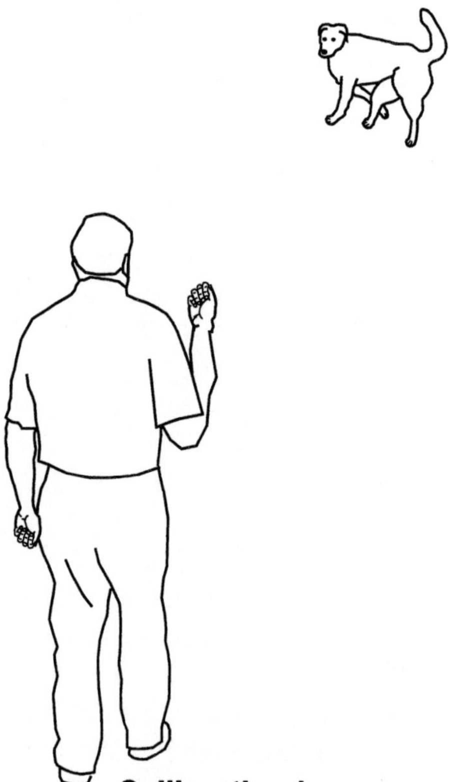

Calling the dog

First we need to teach that obeying the call applies in the park as well. Most likely that will mean teaching him on lead to come when we call him in various parts of any dog walking venues. Recalling a dog is not just as simple as that, but such action goes a long way to proving that power of rank is maintained non-territorially.

Average can confuse

While there can be indications of rank at early puppyhood, it is not until puberty and a little beyond that it will be much more obvious. Higher ranking and dominant dogs are easier to spot quickly. Their pushy behaviour gives the game away. Those of average rank will probably display some slightly confusing signs and need observing for a longer period of time while the low ranker's submissive nature is usually quite clear.

When we determine the rank of a dog that we need to interact with, we also have to consider our own rank. As already stated we are little different in that our rank is also pre-determined. Therefore we need to honestly analyse our own character and determine just where we would fit on that scale. It is difficult to self analyse, as others often see us differently. One can only be as honest as possible, for no one else can work out our score in relation to the dogs unless they fully understand where various dogs would be placed.

I have always given myself a score of 70 which means that I will naturally out rank most dogs but that will not be the case with all of them. My former high ranking bitch I had given a score 9 points in excess of mine. I recognised just how highly she ranked at seven weeks of age and therefore made sure that her perception of my importance was greater than it actually was. To start with it was just a case of ensuring that at all times my will prevailed over hers.

Six months as leader

How this worked out over her lifetime was that, at roughly six month intervals, small signs became apparent that she had decided that I was not as an important member of the pack as she had previously thought. She was trained for competition work and had an immediate fast return when called in a certain manner. Often the first signs were that she would not obey the call within the normal split second but carry on sniffing for another second and a half. It was tempting to let such minute lapses go but I knew that if I did, the delay in response might well double on the following day and continue to double until she worked out that I was definitely of lower rank and that my power when she was free at a distance was virtually zero.

Really Mr Nice guy

My reaction was never to repeat the command for that is the start of teaching disobedience and, of course, this exacerbates the problem anyway. Instead I would run to her and, because her rank was no higher, when she saw me arriving she would show signs of submission but I would still growl at her and show my teeth. In other words I acted like a dog whose ranking order was higher than hers. Within a split second I would back away and then repeat the disobeyed command and over a distance of two yards she would literally hit my legs in her speed to comply. Then, switching to Mr. Nice Guy, I would tell her how clever she was.

Mr. Nice Guy

That would last for roughly six months, again believing I truly outranked her and all would be well until the next time. It must also be remembered that having chastised her I needed to quickly balance my behaviour towards her by finding an excuse to be supportive again. Of course the Mr. Nice Guy act achieved that principle as well.

Over the top

With my low ranking male scoring only 35 such a corrective display of body language and voice would have been way over the top. When I wanted him to do something or desist from an activity all I had to do was ask in a soft voice and he would comply.

As two opposites they have been classic examples of high and low rank, and they serve to illustrate how levels of response to dog's actions need to be adjusted in accordance with their rank. What has to be remembered was while the bitch was extremely high ranking she was not truly dominant. It is not only psychology that changes a dog's perception of our rank, for any practical work which has the effect of manipulating the dog must also allow us to score brownie points. When not allowing my 79 ranking bitch to get away with disobedience, practically, I effectively scored quite heavily. On the other hand, just carrying out basic training will add to the dog's perception of our superiority but it will not necessarily be sufficient to do the job. With low and average rankers it probably will, but as dogs are classified with higher scores, applying psychology will also be necessary in varying degrees.

Other characteristics

As already ascertained, when analysing rank we cannot ignore the other factors of the dog's make up. Therefore we must be aware of any levels of fear or boldness. Insecurity levels also have a part to play in building a mental picture of the dog's nature, and that aspect of a dog's character is dealt with in the next chapter.

BEING SEEN AS THE LEADER

In the wild when the pack goes out they follow the leader. That leader does not wait for pack members who are messing about. It is up to them for they have to learn that survival in the wild is more likely with strength of numbers. When we take our dog for a lead free walk we often inadvertently turn this instinct on its head by following our dog instead. Therefore in non-territorial situations we are, once again, explaining that our rank, and therefore power, is diminished from that apparent in territorial areas.

Keeping them aware

Even if my dogs are foraging ahead of me I will often deliberately turn back or take another direction without telling them. In other words they need to be aware at all times of what their leader is doing for fear of losing him. Also I have to fool them that I am not concerned about what they are doing or the direction they are taking.

Leadership is not all about authority for where applicable we must support, praise or reward our dogs for jobs well done. A dog should never be frightened of his leader but should show respect and what we might assume to be devotion.

For me, the word devotion would be better termed dependency as I believe that what we might term displays of affection are not actually that, but rather displays of their dependency upon us. In turn, and when appropriate, we need to show our respect and support.

When writing this type of book it is possible for readers to believe that, with my own dogs, I am just authoritarian. That is certainly not the case, for I have loved every one of them dearly and spend far more time playing with them and displaying affection towards them than expecting compliance to my wishes. After all, they are my friends. Why else would I give myself the burden of other beings to look after?

RANK AND PECKING ORDER

In the wild naturally high ranking dogs will be at the top end while those born to be followers will be at the opposite end. In the middle will be average ranked dogs. If there are any dominant dogs available then they will be at the very top end. If there is more than one then they will most likely fight for final position. However, that fight is unlikely to be to the death for the intention is only to prove superiority. Once deference is shown by the loser then aggression will cease for their point will have been proved. That applies to males but two high ranking or dominant females may well fight to the death, particularly when one cannot run away. I have known several such instances when two bitches have been confined to the rear end of a car.

Manipulation

Looking at the ranking scale again, as it descends the desire to fight to prove a point diminishes and this is no different with Homo sapiens. Therefore the big question is how is pecking order in the pack determined when lower ranking dogs do not fight for superiority. It really is quite simple for they work out who can manipulate who, and that becomes the other criteria for establishing who is a superior and/or inferior pack member.

Dogs are quite simple, with their limited brain power they cannot think of over a thousand things in a day as we can. Their thoughts will be based largely on survival issues and a few other extras such as companionship and play. When they live with us they have no concern about many of the things we value such as household possessions or clothing as they are issues for our society which is beholden to money.

Snubbing

Some of their emotions are similar to ours as are their reactions to emotions. For instance, when annoyed or wishing to prove superiority they do not necessarily start fighting. Like us they have lower level ways of dealing with such problems. Many of us enact this lower level instinctively as do dogs. The difference is that we know we should not do it as it is the opposite of social harmony. It is called snubbing or blanking out another person.

If you analyse a human snub you will find three main parts to it. The first and major part is a refusal to make eye contact. The second is to avoid conversation, while finally touch contact is definitely not wanted. That might be avoidance of a friendly embrace or handshake. If the avoidance of eye contact is successful then usually the other two do not need to come into play anyway.

Discovering snubbing

Dogs use this technique as a first level of put down. It is designed to send a message to the other that they are superior and that they cannot be manipulated by the one trying to interact. I discovered this some years ago when my 79 ranking bitch and 35 ranking dog ran around me having been playing a game of chase for some time. She was the older and clearly did not have sufficient energy or desire to continue the game. He approached her, eyes bright, tail wagging furiously, with all of his body language saying "Go on run again and I will chase".

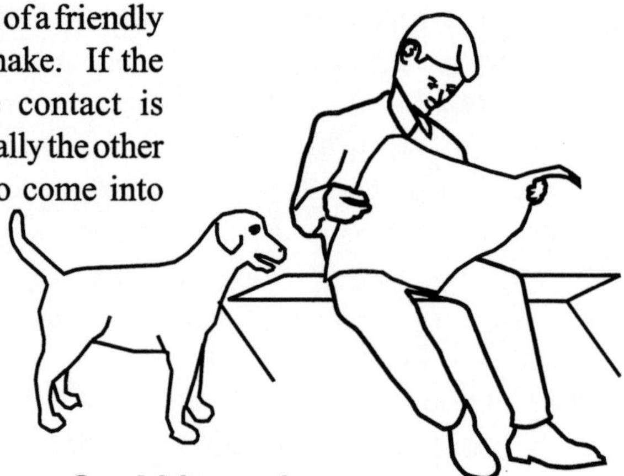

Snubbing a dog

Her response was to move away by about two metres and turn her back on him. Thus there was no eye contact, no touch contact and no vocal contact. He immediately understood her snubbing message and within 5 seconds that lower ranking, very excited dog was totally flat.

Manipulation again

When our dogs try to get us to interact they are attempting to manipulate. It may be that they just want a pat or fuss. Maybe it is a playmate being sought or just to have our attention. What we must realise is that when we comply with this instruction we are being manipulated. If when the dog looks at a door for it to be opened or indicates he wants food or just attention, or we throw the toy he dumps on us, when we comply he has successfully manipulated.

Over a period of time this constant compliance with his instructions can teach him that he is the superior in the pack. I say can, as little in behaviour is straight forward, and it will depend on his rank and his perception of the rank of the person he is interacting with. If there is a large gap in rank with the dog being the inferior to constantly obey will do little damage. If the gap is small with either canine or human being the superior the more likely continual manipulation will eventually give him the perception of superior pack status. If added to other signs of inferiority already evident, the higher ranking dog will see such constant obedience as clarification of superiority.

Privilege of rank

The position between the two may well appear to be confused when he will often still obey instructions given in territorial situations and sometimes when free non-territorially, but he will comply when he has nothing better to do. However, when something is of great importance to him he will disregard any instruction as superfluous, expecting that you should understand his perception of higher rank grants him such privileges. After all, the owner may have constantly displayed pack inferiority by allowing the dog to often manipulate.

Constant obedience to a dog's instructions is more likely to be carried out by a lady. This will be because she can be driven by her maternal instinct to love, nurture and protect anything which is seemingly small and defenceless. That often includes animals. Therefore she will believe she has got it right to follow that instinct with animals. Regrettably average to higher ranking dogs will see this as weakness and not the love of a mother as we might.

It's his voice

Many ladies tell me that their husband's control of the dog is far greater than their own. They will often then clarify that by saying that it is because his voice has a deeper tone. Regrettably I have to tell them that a male's reaction to constant dog manipulation attempts is more often negative for he naturally has less tolerance to being messed about. Let me assure lady readers that males are not perfect dog trainers either. They often also have a dog interaction problem.. We males are often too proud to act in a way interesting to dogs, while the ladies do this much more easily. So both genders have characteristics to overcome before they make good dog trainers.

Ignore what I have written

I need to qualify this subject of manipulation and snubbing by repeating that many people will be able to ignore much of what I have written on these subjects when their dog's rank may not be high enough to make an appreciable difference. Equally many will need to be aware of such pitfalls because they own a dog of higher rank. This is one good reason why owners should understand where the level of their dog's natural born rank lies. Certainly the behaviourist needs to quickly work out ranking values of family and their dogs as, with behaviour problems, the answers will often determine the course of remedial action to be followed. Using a snubbing technique to a low ranking dog is a waste of time for his score is usually so low that obeying his instructions does not add to his perception of existing rank. When trying to increase a perception of our superiority, as the dog scores higher on the scale then snubbing becomes more valid.

Snubbing a dog

When needing to snub, we can achieve this by observing his attempted manipulation using peripheral vision. That means avoiding the all important eye contact. Then we can fold our arms to avoid touch contact and look at something well above his eye level. He will understand this, but if when doing this you feel you are being rotten to your dog there is a way out of this dilemma. He will soon give up his unsuccessful quest and then you can turn the tables on him by calling him to you to initiate exactly what you think he previously wanted. In such a way the owner has become the manipulator. With the majority of dogs it usually only takes around a minute or less for them to realise their attempt at manipulation has failed as they have very short attention spans. If a dog has the habit of jumping up or putting his feet on you when you are seated you may have to touch him. Do so by disdainfully brushing him away but still do not make eye or vocal contact.

Attention now on owner's terms

Dogs definitely understand this body language, receiving a strong message when it occurs. Doing this once will not be sufficient with the higher ranking dogs, but over a period of time it will have made inroads into his perception of your own rank in relation to his.

I cannot stress sufficiently how important full understanding of rank is. Suffice it to say that I believe it to be a major part of understanding dogs.

5

DRIVEN AND APPLIED MOTIVATION

The ability to motivate any species to do something is a very powerful tool. There are various types of motivation and probably the most usual for us is money and the higher the sum the greater the motivation.

APPLIED MOTIVATION

When people talk about motivation, inevitably they are talking about using something such as food or toys as an aid to teaching the dog. If used effectively such motivation usually brings about a dog that will complete a taught training exercise in a happy manner. For me that is applied motivation simply because it is the trainer that applies the motivation at an appropriate moment every time the exercise is practised. If that is so we can also describe the use of something the dog is averse to as motivation to desist from a certain behaviour. Thus in Parts 2 and 3 I may use the word in this book to describe both positive and negative applied motivation on behalf of the practitioner or handler.

DRIVEN MOTIVATION

I also use the word to describe the dog indulging in self motivation. In this case he is not aware of what is making him carry out this action or that it is motivation from within. It is his senses triggering one or more of his instincts, or previously learned behaviours forcing him to do something that pleases. This self motivation I call driven, for, unless taught otherwise, often he does not have much control over his actions.

Such motivation will be apparent in varying degrees within the dog and from dog to dog. It will of course also be affected by years of breeding certain strains of dog to carry out a specific job.

It might be that we want the dog to do something that is a driven motivation so we use an applied version to intensify or refine a sense or instinct, but during this process the dog is distracted by a different driven motivation. At this point the degree of how strong each of these two motivations is will determine which one prevails.

MOTIVATIONAL INFLUENCES

Motivational scale

I like to give motivations a grade on a scale of one to ten for in such a way it is easy to understand why sometimes an applied motivation for behaviour change has little success. If, while carrying out a recall to food held in the hand with a very greedy dog we could say that the motivation scores at least nine. If while doing this a rabbit runs within the eyesight of that dog that was bred to hunt, then there is the confliction of driven motivation brought about by instinct. If that hunting and chasing instinct is really strong then the likelihood is that chasing the rabbit is more important and consequently the driven motivation scores ten. Ten always beats nine, therefore rather than being angry with the dog we need to understand the values of any type of positive motivation for the dog.

Teach the dog to play

If he has little interest in the rabbit then that particular driven motivation will score far less and the applied motivation will triumph. This principle can therefore be suitable to use with any conflict between applied and driven motivations. Of course the dog or puppy that has little interest in food or toys is going to be very difficult to train using applied motivation. That is why we should always teach our puppies to learn to play and thoroughly enjoy it. Even though owning a naturally playful breeds that desire to play should be fostered so that we can ensure that play has a high motivational score. In such cases driven motivation is enhanced by the applied version.

Using food for applied motivation we need to ensure that a puppy is keen on eating and does not finish up as a dog that only eats to survive. While some dogs are born extremely greedy, those with little desire for food, other than enough to feed the survival instinct, are at the opposite end of the scale. We can help encourage some greed by only training the dog when he is hungry and not pandering to his picky mealtime demands until his only motivation to eat is when we place fillet steak in front of him.

Food value

There are also differing treats that can be given to the dog which can influence the degree of greed that the dog has. For example plain dog biscuits have a lowish score for many dogs while meatier foods might well score higher and cooked chicken is a good example. Then there is cheese which many dogs love. Another alternative is liver cake which is usually made to a recipe and beloved by many dog training clubs. Overall we can talk about the value of treats used in conjunction with the dog's greed or lack of it.

Clicker training

Motivation is also the basis of a particular training technique known as clicker training. Basically the dog is taught that when he hears a click he has done what was required and a reward will be immediately or subsequently forthcoming. This is not something to try unless properly understood. It has value as a dog teaching method, particularly for handlers teaching competition work to their dogs. Most usually the reward will be food and less often play. However, it is a method based upon motivation, and everything I have stated about driven and applied motivation still applies.

THINKING LIKE A HUMAN

If we think like a dog then one aspect of our new character will be easy to understand. Play and food will be definite high spots in our day. Usually there will only be one other highlight and that will be going for a walk. Not for us scintillating conversation, hobbies or moving pictures and possibly words on a screen. Therefore thinking like a dog clears the mind, and training will be associated as a very pleasant highlight. It will be something we want to do because it always has the good result of food, play or both. It might even be more important than chasing things that move or generally sniffing around.

Our body language

Our human body language is also a great motivator for it will mean so much to the dog. As he cannot converse he will spend much time watching what it means to him. He will associate differing meanings at different times of the day, but will always know when that language indicates something to his advantage.

Putting on the dog walking shoes, the collar or lead, are often messages expressed by body language. Picking up coats, turning lights on and off, locking doors, opening the food cupboard all convey messages. Maybe we do not want the dog to pick up on certain parts of this language so if we deliberately become inconsistent then the dog can become less certain of the meaning or even lose it altogether. For instance if he starts barking with excitement as soon as the lead is produced we can pick that lead up so many times a day that its meaning becomes very blurred and the trigger to bark diminishes. Better still that the problem never occured in the first place, for we would have been inadvertently teaching inappropriate behaviour by our own unconscious consistent habits.

Deliberate signals

We can also use body language to our advantage, but to do so we must consciously be consistent. Usually we will call these types of our language deliberate signals, such as using our arms and hands to indicate that the dog should come to us. Holding up a flat hand in front of him to specify he should stay where he is. Pointing to our side to convey that is the place we want him to walk.

When relaxing I use my fingers waggled in front of my Labrador bitch to tell her, when necessary, that I do not want to interact with her. She understands what it means and moves away from me although nothing has been spoken. Of course the meaning of all signals must be taught for the dog is not a mind reader even though we might like to think so.

When training or rehabilitating a dog, just the way we move or use certain parts of our body, or maintain a body posture can be a wonderful teaching aid although they may not be deliberate signals as such. We can also back up that language with words if appropriate. When used skilfully our body language can be part of motivation or convey to the dog that motivation will be applied shortly or at the end of the exercise, albeit using food, play or both.

Our habits

Our habits are linked to body language for dogs learn about our habits and often they are an explanation to the dog of what we are likely to do next. They cannot understand other than the sounds and signals we teach them and cannot ask questions. Instead they will soon pick up on our habits to see what messages they convey to them. They will guess why we move from one room to another or check why we are doing so in case it means something different. They can also understand what different clothing we wear means to them. With my dogs the more casual the clothes the happier the reaction. If they are associated with me being outdoors with them they score highly in the ferocity of tail wagging. On the other hand wearing a lounge suit brings forth little joyful reaction for obvious reasons.

Checking clothes

Consistency

Most parents understand that when teaching their children they must be consistent. In fact they can get away with some inconsistency by explanation. That might be allowing a certain thing to happen because it is a special day when normally that activity would not be allowed. With dog training, such inconsistency cannot be understood at all. However, if one aims for 100% consistency and in fact the achievement of that aim is in the low nineties then that will usually be enough for the dog to quickly understand. He can still learn as the consistency level drops further but it becomes harder for him to do so and obviously it can take longer.

Teaching disobedience

Teaching disobedience to a dog is one of the most common traps a dog owner can fall into. This is done by continually repeating commands even though they are disobeyed by the dog.

When we give a child an instruction and it is not obeyed, we usually repeat it, and often keep repeating it. Sometimes repeating the command works with a child, particularly if a harder edge is placed upon it the second or third time. That can also work for a dog but, like the child, there comes a point where he will not obey until the harder edge is apparent.

Repeating commands

On the other hand if the dog is free in the park and we get no response to our call to return our likely reaction is to repeat it. We have already established that the dog learns about our habits so therefore by habit we are teaching disobedience to the command that has not been obeyed. Probably the dog continues with its sniffing in the knowledge that our power has diminished outdoors. Again and again we repeat the command and inadvertently teach the dog that is all that we do rather than insist that our will prevails. That brings about a greater level of disobedience.

Once a teacher in a classroom allows the children to be disobedient over one issue then psychologically there is a greater possibility they will try and win over another issue. Disobedience can then escalate until the teacher cannot control the class at all. It is no different with dogs, where if the leader is not seen to have power, the dog will exploit that weakness. The degree to which this occurs will be determined by the dog's rank. Therefore the leader has to be seen to retain power, but we must remember that power must always be backed up with kindness, consideration and reward just as it should with a good school teacher. To be firm but kind is to earn respect.

REWARDS

Vocal and physical reward

More a traditional reward but still very valid is when we say "Good Dog" etc., and also when we give the dog a pat or stroke that reinforces the vocal reward. Timing is paramount for it should usually be instantaneous with the act which we wish the dog to learn. It might be something we deliberately teach or something we want the dog to desist from doing.

A dog under training for competition will be so used to this type of reward that it can still be effective when delayed a little. However, the household pet dog that has not had much teaching, or indeed with the proposed competition dog that is just starting his education, timing is more critical. It is because the pet dog is inexperienced at the learning process whereas the experienced dog already knows what normally brings forth a reward. That allows a small delay which might be deliberate or out of necessity. Vocal and physical rewards can also be used in conjunction with food and/or play rewards, but to use all four at more or less the same time is somewhat over the top. Remember the principle of too much praise becomes boring. That can be applied to all rewards except for the very greedy or play happy dog which cannot ever get enough of either of these two types of applied motivation.

6

FEAR AND INSECURITY

There is often a connection between dogs that are fearful of sudden loud noises and general over average levels of fear, after all sound is just something else to be frightened of. However, for ease of explanation I deal with it as a separate issue, for not all dogs who are fearful of other things will be frightened of sudden or specific types of sound, while the opposite can also be the case.

SOUND SENSITIVITY

Dogs have good hearing and we are sure that they can hear higher pitched sounds than we can. Generally speaking they can be carefully accustomed to any sound if they hear it frequently enough, as long as the volume has not been increased too soon or the sound is not accompanied by another factor that is frightening.

When dealing with any behaviour problem it is rare to be able to see things simply as black and white for invariably there are always grey areas to take into account. This is one reason why I like to use scales to describe levels of intensities. While there are often other factors that make it harder to distinguish between black and white, levels of intensities could be likened from white through grey and into black. However, for me, sound sensitivity has a simple scale of one to ten.

Scale explanation

Note I have not started at nought for I feel it is unlikely that any dog can be totally impervious to sudden sound of any pitch at any level. The figure one describes a dog that is pretty well bomb proof. In other words fear will not be displayed despite being subjected to very high levels of various noises. Five would be average levels while ten describes a dog with extreme levels of fear. At such a high score a dog will spook at the slightest sound and, of course, display awareness of many sounds before we are capable of doing so.

Panic

He may immediately become mildly or greater stressed to the noise as it becomes closer or louder and may quickly descend into panic. This can result in him bolting, so if free to do so he will run and run with no regard for his safety for he is running in panic from the danger that sound means to him. His one aim will be to get as far away as possible from the sound. If that means running across a busy road his panic will mean he has no heed of his own safety.

Confined in the house he will desperately try to get into the smallest area, often trying to dig while he does so.

Acclimatising dogs to certain sounds as a precaution against developing fear of them is dealt with in Part 2, Chapter 8. This includes remedial therapy with dogs that have already succumbed to this problem.

Steady to gunshot

Many behaviourists have been able to demonstrate both preventative and curative treatment, but the higher the sound sensitivity level the greater the problem and the more difficult the cure.

Continuous therapy

There is also the question of the sound sensitive dog initially being considerably or totally helped, but whose fear of sound eventually returns or worsens. When this happens it does not mean that the therapy has failed but rather that it has not been continuously applied and the result has faded.

Any competition dog trainer will tell you that the quality of a taught behaviour gradually deteriorates if it is not practised with reasonable frequency. That is not to say that intensity of initial training must be continued at the original level to maintain the degree of desensitisation already taught to the dog, but certainly refresher lessons, albeit intermittent, will be required.

A complication to all of this is that some dogs can be taught not to be frightened of certain sounds but will remain frightened of similar ones. An example of this is where gundogs are regularly worked with the gun it might mean they are impervious to the sound of a shot gun being fired but not to the sound of a rifle, pistol or thunder. This means that dogs frightened of bangs might need to be desensitised to various sounds that may be encountered.

Then there is the problem of carrying out desensitisation in a variety of places or venues (circumstantial learning). If the therapy is only carried out in one place, say a room of the house, there is the danger that the dog will only feel safe from that sound in that one place. That is the bad news, but the good news is that the length of time taken to teach the dog in subsequent places should not be as long as in the first place. The time taken depends upon intelligence levels.

SECURITY AND INSECURITY

Awareness of a dog's security and insecurity factors is of great importance to anyone who wants to understand the behaviour of dogs. While we, as the dominant species, also have this as part of our make up, in dogs it is far greater. They were designed to be far more gregarious than us as there is greater safety in numbers. While that also applies to us, the majority of humans are also more comfortable being on their own than dogs. That statement does not necessarily mean that I fully and unconditionally subscribe to dogs being given unlimited human company, for I do not.

Care with advice

When those associated professionally with dogs pass on such ideas, sometimes they are guilty of bringing about suffering for a specific type of dog. They are not aware of this, for if they were they would be more careful with their advice. What I am referring to is the very common behavioural problem of separation anxiety, and that title is very descriptive. It refers to a dog who, when separated from company that he relies upon, becomes stressed. Such stress can grow to be extreme, which then becomes panic. As I am generalising on the subject of a dog's insecurity factors this problem behaviour is not being dealt with here, but far greater detail can be found in Part 2, Chapter 6.

While a pregnant bitch may only produce one or two puppies, this is not the norm as the numbers in a litter are usually far greater. This means that a puppy starts life with a lot of company and this hones his instincts to be part of a pack. His pack instinct also tells him that he will have a better and safer life as part of the pack. I have already discussed this in greater detail earlier in the book under the heading of instincts.

This herding together instinct is of great help to us in domesticating the dog for he will treat us as substitute canines. In many cases he will be happier when the whole pack is together. However just one member of the pack, usually human, providing company on a long or short term basis can bring about peace of mind in the dog. How happy or unhappy he is about pack company will greatly depend upon the strength or weakness of his security/ insecurity factor.

I developed another scale to describe levels of security or insecurity. Once again I do not start at naught as I cannot believe any dog has total or absolute security when left alone in a strange place. Details of this scale can be found in Chapter 6 of Part 2. It is in that part of this book as levels of insecurity have a great bearing on the behaviour known as separation anxiety.

Mother and puppies

High scoring example

I have a delightful Border Collie male called Whizz. His characteristics as applied to insecurity are high with a score of 10½. He cannot get enough touch; where possible he follows us around the house. He will not stray further than 40 yards when distracted, but for him that is an extreme distance, therefore he will quickly reduce it. These characteristics are typical of a very insecure dog. He is a prime candidate for separation anxiety yet he does not have it and I am sure that it is my understanding of the problem that keeps this stressful temporary mental illness from afflicting him.

Low scoring example

On the other hand I have owned a Border Collie bitch to whom I gave a score of 1½. Her characteristics which brought about such a low score were that she had little interest in being touched, stroked etc. In fact she never sought it and, if it was initiated by others, about five seconds was her tolerance. She never had any desire to follow my wife or I around the house or garden. If we went out for the evening she would briefly open one eye to note that something was happening and then go back to sleep.

On a lead free walk she was quite happy to stray many hundred yards from me. In other words she was almost totally without the need for back up. This may sound like a boring dog to own but in fact she was a lovely dog and, thankfully, she was well trained, so recalls from several hundred yards were not a problem. These characteristics are typical of a dog with very low levels of insecurity.

Other factors

As with all behaviours there are other influences that can make what appears to be a straightforward diagnosis a little complicated. I have already said that fear and rank can play a part, but so can the difference of venue.

A dog taken on the same walk and same route every day will display slightly less insecurity should his owner move briefly out of sight. On the other hand that same dog in a walking venue he has never been to before will display slightly higher insecurity levels than normal if a similar situation occurs. Therefore all influencing factors must be taken into account before scoring a dog.

FEAR

The one characteristic I probably see more of than any other is fear. I suppose that would apply to many animals, and we must always remember the size of dogs compared with us. Of course insecurity brings about fear as does sound sensitivity, but as they are specific fears they are better separated from general fear. However all fear can start as stress and finish up as panic.

Latent fear

Fear is quite a complex subject. It is my belief that all fear in dogs is inherited and I sometimes refer to this as latent fear. What I mean is that even though fear may be a learned behaviour it must have been latent through inheritance, albeit to a small or greater degree.

A puppy whose fear is very obvious will not require a learned experience from a fearful stimulus to bring it out. Therefore I believe that genetics, combined with the interference of nature, are the origins of all fear. It is the latter factor that would account for differences of fear levels between puppies in the same litter.

Fear triggers

Fear is triggered as a result of a dog's perception of an immediate threat or danger in the environment. Common perceptions are people or children, and these are sometimes further split by gender. Another fear can be triggered by the proximity of an unknown dog.

Such fears may be breed, colour or size specific. Other animals, sudden noises and sudden movement can trigger fear. Traffic, machinery, domestic appliances or passing an area where a threat has previously been a perception can be other examples fear becoming apparent. Fear of travelling by car or the approach of others towards the car the dog is in can also be a problem.

While socialisation carried out in a controlled manner to that which is perceived to be a threat can often bring results, dogs with extreme fear can be difficult to rehabilitate. If a dog is frightened of people and/or dogs and he is withdrawn almost entirely from contact with the outside world then that dog will withdraw into himself even more, only trusting those he lives with. Therefore isolation is the opposite of socialisation.

If remedial therapy is not applied with such a dog there is no possibility that he will "Get over it" or "Grow out of it". In fact such statements made by those without in depth knowledge of the subject do a great deal of harm.

Defensive aggression

Fear is experienced by dogs at differing levels and the perception of how great the threat is will affect those levels. However, a display of the highest level of fearful body language to a stimulus gives us an indication of how bad a problem is.

Just because the dog becomes fear aggressive is not a total indicator of the level of fear, for the dog's natural born rank (see chapter 4) will also play a part. While not a rule set in stone it is far more likely that a very low ranking dog fearful of the approach of others will back away or display immediate submission while the opposite high rankers will more likely attack. Once again there are grey areas for any non violent creature will fight

Fear and submission

if they feel their survival is threatened. In such cases we can determine this as defensive aggression.

Often associated with rescue dogs is the perception that because the dog shows fear of people, particularly when it is displayed against men, he must have been abused by people or a specific gender in the past. While it is always a possibility, such statements are pure guess work and often not the case.

Trust from the nest

My own observations of dogs and people are that is more likely the dog was conceived with sufficient levels of fear already built in for it to subsequently surface.

When mother and litter are solely looked after by a human female, if latent fear of people is part of a litter member's make up, it is far more likely that females will subsequently be trusted. As litter duties more often fall to our females puppies become accustomed to being handled and fondled by them rather than by males.

I have seen fear in all breeds of pedigree dogs and mongrels as well. Sometimes it is caused by indiscriminate breeding and this is why those seeking a puppy should only buy from a reputable breeder. Maybe nature might still interfere but reputable breeders do try to produce litters sound in mind and body.

Rehabilitating from fear is easier with younger dogs and also with dogs that have not shown fear to a certain stimulus for many years. A rescue dog can have been fearful of something for years and only the new owners try to help the dog. This is more difficult, as are the higher levels of fear, but someone trying to help owners with problems should not immediately write a dog off but rather go into matters deeply to see if it is possible to rehabilitate him.

Distraction for fear

I am not a great lover of distraction as a method of combating fear. While it can work, far too often the level of fear is greater than the applied distraction motivation. My experience is that it is far more likely to be successful with mild cases of fear, and when fear is extreme no pleasant motivational factor will distract. Reassuring a fearful dog is another tricky question and I explore this more deeply under the heading of Stress. These days my conclusion is that sometimes it is valid and sometimes it is not. I use all my practical experience with dogs to make that judgement. Without experience such a judgement cannot be made. This is another example of sound practical handling experience being a pre-requisite of solving dog problems.

Not a threat

What is very valid is to do everything possible to teach the dog that whatever he is afraid of is not a threat to his existence. Under the following heading of stress I describe how I enter the house of a client with a problem dog. If that dog is fearful of strangers, one of the many aims of my visit will be to teach the dog that I am not a threat. In other words my visit should be the first lesson of very many in controlled socialisation for that dog. Greater details can be found under that heading.

Likewise, when I have been in the house of a dog extremely

fearful of strangers, which after an hour will still not go anywhere near me, I will have to change my tactics to gain the dog's confidence. Maybe it is a dog which is displaying aggressive signs to any movement of my body while I am sitting, etc. Inevitably I shall need to go out with the dog and owner to establish how we are going to deal with the dog's fear born aggression of passing strangers.

Catch them up

What I ask the owner to do is leave the house before me with the dog on lead and walking roughly to heel. I then leave 20 to 30 metres behind them and very gradually catch them up. Without any vocal or eye contact with owner or dog I just move smoothly in beside the dog so that he is effectively sandwiched between the two of us. If all is well, after a time I quietly ask the owner to transfer the lead to me. Then as we continue walking I subsequently ask the owner to gradually drop behind so that there comes a point when the dog realises he is walking with me and not his owner. Gradually I start speaking to him intermittently in a soft reassuring voice. If he passes that test I then occasionally make eye contact, and sometimes I can eventually follow that up with a reassuring pat and/or prolonged eye contact.

If necessary, during this stage of the walk occasionally my intuition will tell me that I can correct the dog for pulling on the lead without adversely affecting his trust in me. What is remarkable is that when we eventually return to their home and the dog is allowed off the lead his demeanour to me is always markedly far better than before the walk. It will of course vary from dog to dog. Some have completely lost their fear of me by this stage while, at the other end of the scale, they no longer totally avoid being near me but do not voluntarily seek my close company.

Improved immensely

I cannot be certain why this works. I have some ideas but nothing that I would be ready to put in writing at this stage. However, it is quite uncanny to be in a room with a dog that may fearfully growl and even lunge at your slightest movement while restrained by the lead, only to witness that dog's behaviour towards you to have improved immensely when returning to the house with him.

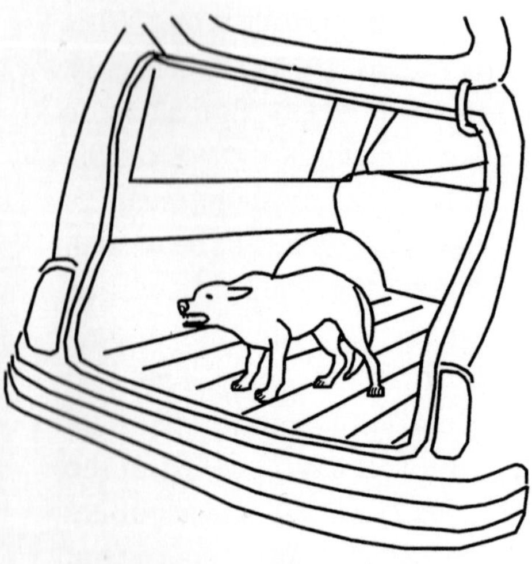

Aggression in car

Of course we do know that when dogs display fear, the smaller the space they are confined to the worse will be their fear. An example of this is a dog confined in a car which is approached by someone. While it is possible that the dog may aggressively bark for territorial reasons, it is quite often fear driven. Why reduced space enhances fear is easy to understand for the dog is denied the chance of flight.

Dogs on lead, fearful of the approach of something will display far greater fear than if they were off lead. Of course safety dictates that they may not be able to be off lead but they will be aware that flight is impossible. When it is a case of fear aggression to another dog then the aggression factor will be greater on lead than off for the same reason.

Behaviour and drugs

Drugs are increasingly being used to combat behaviour problems, in particular those associated with fear.

I am quite happy for drugs to be used for medical problems, but I would far prefer that remedial therapy be fully explored before resorting to pharmacology.

Regrettably, I feel that the future trend will be to use drugs first for it is something that most veterinary surgeons readily understand. Hopefully there will be those who resist the temptation to prescribe medication and first use the service of a behaviourist who has provided results in the past.

STRESS

Stress is a word that found much greater use during the latter part of the previous century. It could almost be described as a fashionable word. When a person complains of suffering from stress it is one of the most difficult things to prove or disprove. Usually for humans it refers to people who are suffering from depression, anxiety, or other neuroses.

Much has been written about stress, and of greater interest to this book is canine stress. Some of what has previously been written is so detailed in the myriad of body language signs a dog can display indicating stress, that dog owners could become obsessed by it. In fact they could become stressed themselves by worrying about stress and their dog. They would display this by watching the dog intently and worrying disproportionately about this subject.

There are those who do not believe that stress exists. While that does not apply to my thinking I feel we need to get it into proportion as, for me, fear must be the general heading that it comes under. In other words I am saying that canine stress can be seen as low levels of fear, and therefore an escalation of stress caused by a particular stimulus would be fear and a further escalation would be panic. That is how I chose the title of this chapter.

Therefore when, in certain situations, we see mild signs of discomfort displayed by the dog, we are looking at fear which at this stage might be better termed stress. What we should do is try to ensure that stress subsides rather than allowing it to escalate into fear.

I do not intend to list all the possible signs of fear for I feel that would confuse most people. Instead they should be able to obviously see signs of discomfort from the dog and act accordingly. How to act will depend upon the place and what the dog and owner are doing at the time. The stressed dog may look to you to observe your reaction and, generally speaking, you should take the attitude of "problem, what problem".

Intuition

I do not fully subscribe to the theory that you should not console the dog by voice and/ or touch for fear of reinforcing his reaction but, rather, would suggest that you should use your intuition in differing circumstances. For example I do not normally support the dog displaying stress to bangs. Instead I might play with him to bring about an association of fun when the noises are occurring.

In other circumstances I might withdraw the dog from the stimulus that is adversely affecting him. In some circumstances I may feel the risk of likely reinforcement of fear is overcome by the possible benefits of soothing. There is no hard and fast rule so I can only give you the basic information for you to make a judgement at the time it occurs.

If the dog is being trained and displays stress then he may be confused or have other worries, and you would need to make an intuitive decision about your reaction. It is possible that you may deliberately be using mild forms of stress to combat greater stress. On the other hand what you are doing might be the last thing you want the dog to have an association of stress with.

Pleasure to stress

A classic example of this for a competition dog trainer would be when teaching heel work to a dog with his tail wagging in pleasure and then his body language changes to that of mild stress. Now this example can cause further confusion in our minds for it is possible to misinterpret some canine concentration of the task in hand with stress.

This reinforces my opening remarks about this subject becoming an obsession. If stress is becoming associated with competition exercises then the training has become too formal with insufficient triggers for happiness such as praise, food, or play. Maybe the trainer should relax or take a break. Maybe they should stop altogether and try again another day but with greater applied motivation to keep the dog happy.

A calming word

If while working with a dog you feel your actions are inducing stress then avoid direct eye contact for that can be interpreted as threatening. There is an argument for teaching the dog a calming word. This he should come to understand as everything is OK. But it should be different to normal praise for a job well done. Additional to this could be the motivational word "Yes" which I add to any training I carry out with my own dogs. It means to the dog that he has got it right and shortly, but not necessarily immediately, a food reward is coming. I always deliver it in an exciting manner. Now if my dog was displaying stress I am not going to say "Yes" for that would reinforce his current stressful demeanour. However if I started to play with him to teach him that he should associate fun with the stressful stimulus and he responded positively, then the word "Yes" would then have an appropriate use.

The three No's

I have mentioned about avoiding eye contact, and now is the time to expand this. If a dog is displaying fear at your approach or at his approach to you, then direct eye contact should be avoided. If two dogs of higher rank, unknown to each other, are displaying direct eye contact then the immediate likely outcome is aggression. Dogs have a perception of certain types and movements of body language, and direct eye contact is one. The outstretched human hand of friendship is another, for the dog fearful of your attentions will see that as threatening, confirming the perceived initial eye contact threat. When you back up both of those with your friendly language obviously directed to the dog, all these actions confirm a possible threat.

Visiting the dog house

Therefore whenever I visit a house to see the owner with a problem dog, initially I do not do any of those three things. My whole first objective is not to do anything that the dog might perceive to be a threat. Within seconds I may judge that I can do those things without invoking stress or fear. On the other hand I might decide that I should continue with this ploy for longer. What I really want to occur is for the dog to check me out while I totally ignore him, instead keeping my eye contact on his owner. I want him to find me entirely non-threatening and, in the case of an excitable dog, boring.

Ignoring the dog upon entry

If he is dominant I do not want to interact with him in any way, including eye contact, for that can trigger aggression in such a dog which feels challenged by the eyes.

Aggression problem

In this way I have crossed the thresholds of thousands of houses and have never been bitten or attacked in any manner while doing so. I have no worry about entering the houses of the most fearful aggressive or dominant aggressive dog, and to carry out my work effectively I must be able to display a relaxed attitude when in the company of any dog.

Do not think I do this under any circumstances, for if the initial telephone contact from the owner suggests an aggression problem with callers at the house I may ask for the dog to be initially contained in another room. However, I have never ever left a house without subsequently being in the same room or area as the resident dog.

PANIC

I have already stated that there are three main levels of fear, and panic is the word associated with extreme fear. When a dog panics, all reason goes completely and the most highly trained dog will disobey an order, instead obeying the emotions of his panic. I have witnessed a highly trained competition dog bolting in absolute panic for five miles without pause, caused by an extremely loud car backfire. That normally highly trained dog was oblivious to any commands to stop. After five miles of flight his panic had subsided sufficiently, or his energy totally expired to allow him to stop.

Air brakes

I have also witnessed a dog sitting obediently on lead at a kerb with his owner waiting to cross the road. A coach pulled up as the traffic ahead slowed and immediately within a few centimetres of his nose the coach's airbrakes made their loud hiss right in his face. His panic was total to the extent that the owner could not stop his attempts to slip his collar. He eventually did so and by then the traffic was moving, so straight under the wheels of a car he ran. Luckily he was a big dog and the small car was only just gathering speed, so rather than being run over he was pushed along the ground for several metres until the motorist stopped the car. Bruised and lacerated he spent a few days in veterinary hospital but luckily lived.

When a dog is extremely destructive having been left alone in a house, his state of separation anxiety has driven him to panic and his attempts at release from that panic is destruction.

Of course we know that bowel and bladder release can be a result of panic in humans and it can be the same with canines.

Buried in junk

I have also seen a dog try to bury himself in junk under the stairs during a thunder storm. My advice at the time was to let him do it until the extreme fear subsides. All of these examples are panic where reason cannot prevail. Generaly speaking there is not much we can do to bring a dog back from panic other than to remove from him, or remove him from, whatever is causing the panic.

For fear of reinforcing panic, to soothe or console is debatable and, while there may be arguments for doing this with lower levels of fear, I feel

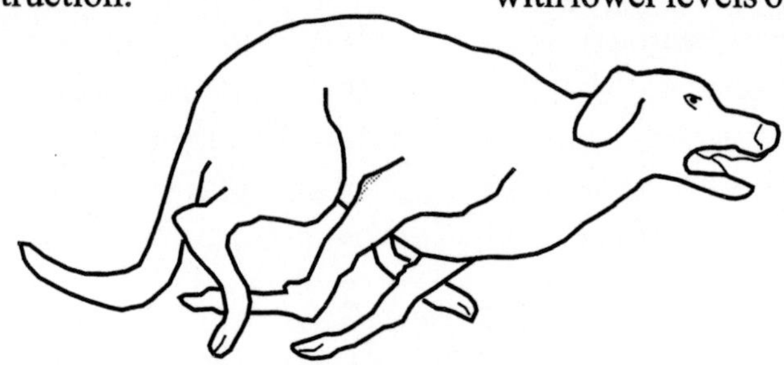

Panic

that it is far better to let the dog subside from panic through to fear and then when perhaps still stressed, albeit at much lower levels, to consider reassuring him at that stage.

Conventional wisdom says there are no circumstances where consolation should be given, but having carried out experiments with differing dogs I do not find that to be applicable to them all. Therefore intuitive decisions should be made. Yet again intuitive decisions are examples of the necessity of sound dog handling experience when advising people with problem dogs.

7

CONTROLLED SOCIALISATION

This chapter has the deliberate title of controlled socialisation for the emphasis should be on controlling just how the socialisation will take place. Many people think that socialisation is purely a means of introducing the dog to what he will meet in everyday life. Of course doing so is right but it must be carried out in a controlled manner.

SOCIALISATION EXAMPLES

The following are examples of risky socialisation. A dog fearful of people making direct approaches to that dog will be having a reinforced lesson in that fear when people do so. A puppy running free in a socialisation class of other free puppies who are supposed to be learning canine social behaviour can be taught the opposite of what is trying to be achieved.

It may be that another puppy is being aggressive to it, thereby inducing fear and/or an equally aggressive response. When a puppy's first lesson about vehicles is close to heavy moving fast traffic expect him to be frightened. Taking the puppy to the field where the horse charges at it. All these are examples of socialisation attempts that will probably achieve the opposite. There are many more which common sense should dictate.

When we talk about socialisation, what we mean is accustoming the puppy or dog to all the facets of human life that he might be exposed to so that none become a stimulus that induces fears. Strictly speaking there is another word associated with socialisation and that is habituation but that is getting into jargon so, for our purposes, all will come under the one heading of socialisation for that is what most people understand.

When discussing dogs it was only in the latter part of the twentieth century that the word socialisation came into common usage. It would appear that, prior to those recent times, socialisation of dogs was unnecessary. While I am sure that cannot entirely have been the case, at one time accustoming dogs to every day life in a house and community seemed to occur more naturally. Now it is something all dog owners need to consider as something essential.

Animals, people and sounds

Canine socialisation is really a process where the animal learns how to calmly recognise the species it meets. This may be man, cats, dogs or other animals. It learns how to interact with them and develop communication skills where appropriate. Accustoming the dog to other sights sounds and smells he may have to contend with during his life are just as important. For instance the kitchen utensil that falls, the roar of heavy traffic, the noise of building works, thunder, fireworks and so many other things need to occur so that at no time does he feel threatened.

Ongoing socialisation

While the most receptive period in a dog's life for learning is from 3 to around 16 weeks, this is not a limit and, if necessary, socialisation should continue beyond this. It will of course depend on each dog, for some will need more than others. There will also be those dogs where, if initially successfully socialised only to subsequently be withdrawn from society, they will revert to the same or greater levels of fear that was latent within them or had already developed. In other words some need ongoing socialisation and, while this might be the dog with greater levels of fear we cannot presume that it will not apply to those with lesser levels of this instinct.

If this is to be a requirement then owners should carefully watch for tell tale signs of fear that might have previously been eradicated. They should never take the attitude that it will go away or what more can they do, for if there is to be any further change it will be for the worse unless remedial action is taken. Dogs can be sexually mature from 6 to 14 months of age. It can differ from breed to breed, within a breed and within a litter. During this time they will often commence barking at strangers and become much more territorial. It is at this time that males first start cocking their legs to let others of their gender know they are around. Introduction to strangers visiting the home becomes important at this time.

A bottom line

It is wrong to only associate this preventative work with puppies, for socialisation is the bottom line of all remedial work with dogs acting in an anti-social manner as a result of fear. I say bottom line for even if the fear induced anti-social behaviour is eradicated, unless successful socialisation to the original fear stimulus is carried out a cure has not been affected. Instead only a screen to fearful reaction has been put in place.

Remember

As fear remedial work with adult dogs will always include socialisation, the following references are for puppies and adult dogs requiring rehabilitation. Above all, socialisation is about teaching the dog that what you introduce him to is not a threat to his existence. He is a creature of habit and what you teach inadvertently or by design is what he is likely to remember.

SOCIALISATION TO OTHER DOGS

Always try to ensure that the other puppies/dogs are not fear aggressive, too highly ranked or boisterous. Allow them to meet one to one at a time on lead and to check each other out. If both display non aggressive body language positive socialisation is taking place. Allow the puppies/dogs to interact naturally but gently curb over exuberance. Try and socialise to a variety of breeds sizes and colours.

Special classes for puppies can be very beneficial for socialisation purposes. However I do not believe that they should be able to do their own thing off lead for the simple reason that there will inevitably be differing characters of canine amongst them. Already higher ranking ones may be over pushy or even aggressive.

Then there will be the fearful ones who need to learn that the others are not a threat. On lead all this can be controlled by the instructor, and that is how I believe it should be. While they need to learn the social skills of the canine world it is far better that they do so in an ordered manner.

Socialisation to other dogs can also take place while on walks, no matter what the age of the dog is. However, avoid any puppy or dog that may induce the instinct of fear in your dog by displaying fear, dominance or over-boisterous behaviour himself. If you do not know the approaching dog/puppy then make a calculated guess.

Dogs socialising

Do not constantly pull him away from other dogs or you will be sending out a message that avoidance of them is paramount. Always ask the owner of the other dog about their dog's demeanour. If the dogs interact naturally sniffing each other or with gentle play signs then let the dog know he has got it right While we know dogs can be more fearful on the lead we have to balance that aspect against retaining control. If the other owner also has lead control so much the better. Do not worry if all the usual sniffing routines are taking place. That is a normal canine social habit.

Body language is easy to observe, either aggressive, calm or playful. Do not allow socialisation to become "Whoopy! Let's have a rough and tumble". Socialisation is not about teaching your dog that all other dogs want to play.

Care with children

Socialisation to just your neighbours dog is not sufficient; neither is it when confined to one or two breeds, or certain sizes of dogs, or to adults with children excluded. However using children for socialisation purposes is difficult, for at all times we have to concern ourselves that a child should not be put in danger.

SOCIALISATION TO PEOPLE

(1) Do not allow people to make a direct approach to your puppy or dog.
(2) That means no immediate eye, hand or voice contact.
(3) Make sure eye contact is the last of the three used for its effects can be devastating.
(4) Allow the puppy/dog to check out the stranger without them immediately indulging in 2 above.
(5) Assess when each of 2 above can gently be achieved without adverse effect, starting with voice.
(6) Ensure an outstretched hand of friendship goes under and not over the puppy's/dog's head.

(7) If food, treats, etc., are used, then do not allow food to be pushed at the puppy/dog.
(8) Instead of 7 above, allow the puppy/dog to find food in the hand seemingly glued to the leg of the stranger.
(9) Try and ensure even balance of male, females and children of both genders.
(10) If your dog lives in a childless house, socialisation to children of differing ages could well be important.
(11) Try and ensure that people visit the house for socialisation to also take place at home.

Further explanation

A direct approach is fine for a perfectly adjusted dog. However, just think how you would feel if a stranger virtually pounced on you and ruffled your hair while looking you directly in the eyes.

Remember that eyes, hands and voice from a stranger coming straight at a fearful dog will only reinforce his perception of a threat. That is the antithesis of socialisation. Once the dog/puppy has checked the stranger out by sniffing etc. allow occasional gentle vocal contact first. Do this without at first allowing an outstretched hand. Such a hand can be seen as a threat and not friendship. Finally, occasional, and to start with brief, eye contact can be made.

While, for the purposes of imposing leadership, any interaction should be on our terms, socialisation is the one exception where we should let the initial contact be on the dog/puppy's terms. All of these three types of interaction will, if possible, be at one meeting and the trainer must judge when they should occur.

It is amazing that people, especially dog lovers, do not learn that when attempting to touch a strange dog the hand should go under the chin and not over his head. A latently fearful dog, when losing sight of a hand disappearing over his head has had his fear triggered unnecessarily.

When using food for friendship let the dog go to the treat even if he has to find it by nose or accident but do not allow the food to be pushed towards the dog.

I have already stated that an outstretched arm can be seen as a threat. Ensure a stranger's hand containing food stays stuck to his trouser or legs in this way it is non-threatening. All types of humans must be met without any of them inadvertently or deliberately threatening the dog/puppy. Take care if your dog never sees young children during his formative months; he may subsequently consider their oft erratic behaviour to be a threat. Socialisation needs to take place anywhere the dog might meet strangers.

Fear of man

SOCIALISATION TO OTHER ANIMALS

(1) Always make the introduction from a distance then gradually close it as progress is made. If this means over many sessions, so be it.
(2) If the other animal bolts this can trigger fear or predatory instincts in your dog so try and avoid it.
(3) Many other animals do not want the close socialisation that occurs dog to dog so do not try to enforce it.
(4) Any aggressive signs brought about by fear, dominance, territorial or the predatory instinct must be curbed so that your puppy understands that a human pack superior finds that behaviour to be unacceptable.

SOCIALISATION TO TRAFFIC AND TRAINS

(1) Approach traffic gradually from a distance.
(2) Approach from the tail of a T-junction thus starting quite a distance away.
(3) Otherwise choose another venue or use your initiative to start at a distance.
(4) Do not commence with thundering traffic.
(5) Make haste slowly with more frightening traffic.

SOCIALISATION TO PLANT MACHINERY AND DOMESTIC APPLIANCES

(1) Approach gradually from a distance.
(2) If possible start with light noises.
(3) Consider playing with the dog while noise continues to be emitted.

PRINCIPLES FOR ALL SOCIALISATION

(1) Where safety and control are necessary the puppy/dog should always be on lead.
(2) Once the dog has accepted the target as non-threatening then move on to another.
(3) Continued socialisation to that already accepted as non threatening by the dog is pointless.

(4) Where it is possible occasionally change rooms, places or venues for socialisation.
(5) If any adverse signs are displayed by puppy/dog then revert to a previously successful stage.
(6) Curb at an appropriate level any aggressive signs from your puppy/dog.
(7) Be prepared to take days, weeks or months dependant upon the nature of your dog
(8) Be careful about perception of a threat being inadvertently reinforced by reassurance.
(9) Do not allow over socialisation to people or dogs so they are considered as toys.
(10) Very careful outdoor socialisation should be started with the puppy carried in the arms.
(11) Once socialised successfully, a puppy/dog should not be withdrawn from society for fear of reversion.
(12) At the first sign of reversion, re-socialise and then maintain occasionally.

FLOODING

This word is quite descriptive in that it describes a process where the dog is flooded with what he perceives to be frightening as a method of desensitising him to it. However, it is fraught with danger for those without knowledge of carrying it out. A person with in depth knowledge of remedial behavioural techniques needs to make a decision as to whether the dog could benefit from this technique or whether the opposite might be the case. Also the degree of control applied to the process will vary from case to case and, again, making that decision requires expertise. Therefore forewarned is forearmed.

Flooding is usually used as part of a programme to overcome general fear problems and in particular sound sensitivity. Refer to Chapter 8 in Part 2 for details about controlled flooding.

8

AVERSION THERAPIES AND CORRECTION

It is quite interesting that the academics gave us the term of aversion therapies and it was not that long ago. While the word correction became unfashionable, in came the more modern terminology. However, they still mean correction, for an aversion therapy is something which is applied that the dog is adverse to. I am sure dogs are averse to any forms of correction and some aversion techniques are not new. A classic example of this is noise aversion therapy, for at least thirty years before the term had been coined for dog behaviour some dog trainers were using noise as a corrective technique.

Correction or aversion

There is a slight difference in the two terms, for correction has a connotation of a being physical while aversion therapies are usually more remote.

When I say physical I refer to using the hands to help the dog to sit, using the lead to check the dog on the collar, or something similar. While on that subject I must refute those who believe that to check a dog in this manner is cruel, injurious or dangerous. It is all a question of how it is applied. I have found that those who think this method should not be used are usually incapable of skilfully applying it anyway.

No universal method

When dogs have certain behavioural problems such as aggression and are not responding in any other way then correction is often the best way to stop it. Part of the skill is to be able to ascertain the most suitable form of correction and the degree to which it is delivered. If successful that does not mean that rehabilitation is complete but usually that a major step has been made along the road to rehabilitation. If distracting the dog from the stimulus has been a total failure because his driven motivation for aggression is greater than the applied motivation, then correction might well be the answer. As no one method of applied motivation is successful with all dogs, neither is one method of behaviour modification or indeed correction. Always the skill is in choosing the right one.

Dog to dog aggression

APPROPRIATE LEVELS

In this day and age correction has to some extent become a dirty word among those succumbing to the fashion of never correcting child or dog. In this book I have stated many things. Some are: how it is possible to teach by positive motivation only, how a dog trainer must never be cruel, how a dog needs kindness and support. I have also stated that I do not believe that discipline can be discarded. My reasons are that if we believe we should understand dogs then we should know that a wild pack of dogs will have discipline imposed by their superiors. The degree of discipline used will usually be commensurate with the degree of stepping out of line. Those leaders work entirely by the instinct given to high ranking or dominant dogs and I wonder why they, like man, were given such an instinct if it is wrong to use it. Someone has to be a leader in all walks of life albeit human or canine.

I have deliberately left this chapter until the end of this part of the book for, when using corrective techniques, what is of paramount importance is to absolutely understand the character of the dog. Yet again I come back to this facet of dog training knowledge which is only gained by a combination of theory and much practical experience. First the rank of the dog must be established, and then levels of fear and insecurity. If noise is to be used then levels of sound sensitivity need to be understood. Was one of his instincts in confliction with an instruction we have given? There are so many factors to take into account.

With a combination of these factors a decision can be made as to what is an appropriate level of an aversion therapy or correction to be applied. To just go blundering in without such knowledge is to either over or under apply behaviour modification tools. We might just be talking about vocal admonishment and therefore will the dog cringe if we deliver too strong an "Arrgh". On the other hand will he completely cock a snoot at a quietly issued vocal sanction? On this occasion can we forget the misdemeanour altogether, or will that be setting up a habit that he can get away with something. It isn't easy to make such decisions with a dog, but then being a parent of children isn't easy particularly at the ages they start to challenge authority. It is just the same with aversion therapies.

Becoming a dog

To better understand dogs I try and become a dog mentally. That includes discipline, but naturally I do not use my teeth as dogs can do as an act of correction. However, I can show them my teeth and I can growl in my language if I think it is warranted. If I really think it applicable I can hold them up to my face at the same time, but that would only be with my own dogs. However, that is as far as I go and rarely resort to that sanction anyway. My golden rule is to be seen to be firm but also fair and supportive.

DANGER WITH DOMINANCE

As a dog I must understand that if I accept a challenge from a dominant dog then there is a good chance that his response will be to bite me. He will not be trying to maim or kill but rather to prove a point of superiority. However his bite or bites might inadvertently do more damage than he thought necessary.

What I am saying is, just as stated in my table of ranking values, that no attempt should be made to use aversion therapies or anything practical with a dog of a dominance level that is high enough for him to retaliate or escalate his aggression. For me that is a ranking score of around 84. With that level and above, if I tried to practically dominate such a dog then it is almost certain he will come back at me. Maybe that score will not be right for the practitioner who must work this out for himself. Therefore the conclusion must be that if the behaviourist feels at risk of being bitten when taking the next step, then it is almost certain that advising the owner to indulge in confrontation will bring about the same result.

Aversion therapies and any practical handling techniques contained within this book should not be attempted with a highly dominant dog unless skilled in the art of evaluating dogs and their likely reaction, or the result could be dangerous. If the practitioner does not have sufficient skill to do this they should seek expert advice before taking steps into the unknown.

SNUBBING & TIME OUT

There are other ways a dog can be corrected. Snubbing is a mild form of correction for it is in effect pulling rank and stating by body language that the other is an inferior. Time out, as the Americans call it, is another. Perhaps better termed as corrective isolation it is when, as a correction, the dog is sent to another room to be left alone. With this correction one has to be sure that the dog understands why time out is occurring. I cannot teach that, for one only knows that by instinct born out of experience. Also the dog's level of insecurity will play a part with this sanction as it will have little effect on an almost secure dog but can have a considerable effect on one that is highly insecure. In the latter case that could be deemed to be detrimental to the dog.

APPLICATION AND HABIT

Using the lead to check the collar is a physical correction and when applied skilfully it can be very effective for some dogs. Such corrections, for that is what they are, can be used in varying degrees of severity, but they are usually not appropriate for a puppy or young dog being taught in the first place. Such training can mostly be carried out by positive motivation, however we must remember that driven motivation might outweigh applied positive motivation and the result can only leave the handler with correction as an option. The skill is knowing when and what is necessary and to what degree it should be used.

Immediate application

Whatever correction is used the dog must connect it with his actions. If that does not occur, then correction will only bring about confusion in the dog's mind. That is why timing of corrective action must be immediate with the behaviour we wish the dog to desist from. All dog trainers know that, as explanation is impossible, the only way a learner dog connects the correction with his act is for the correction to be instantaneous.

Simple equation again

Far earlier in this book I wrote about how a dog associates his actions with the simplicity of good or bad result. I prefer to liken that to a simple equation but not a mathematical one. I call it the principle of 'A=B'. Therefore 'A' is his behaviour and 'B' is the result of his behaviour. If 'A' is to be aggressive and consistently he finds that immediately he is aggressive 'B' occurs, and for him 'B' is a bad result and something he would rather did not occur, then he will learn not to indulge in 'A'. If before correction is applied he considers 'A' to be a good result then his behaviour is being reinforced.

If it is necessary I use three main aversion therapies with my work of behaviour modification. These are with cases of dogs that have developed a habit of what is often termed inappropriate behaviour. Of course for the owner it means the dog's behaviour is a real problem to them and possibly others. Often connected with circumstantial learning such unwanted behaviour has sometimes been inadvertently taught by the owner, or is a reinforced instinct or merely something the dog has learned to do. Correctly applied aversion therapies invoke the simple 'A = B' equation.

Habit

One thing that is usually common is that the behaviour will be very much habit and part of the remedy will be to break that habit. That remedy will depend upon the cause of the behaviour and the overall character of the dog. Often the application of the remedy will require very skilled dog handling and experience of dogs and their likely reactions.

As already discussed, almost inevitably psychology will be part of the advised cure but alongside that will be the use of one or more aversion therapies. Where fear is the cause then certainly the final part will be socialisation to that which the dog is frightened of. That must never be forgotten. Fear would have been the cause of the aggression, therefore until the cause has effectively been dealt with full remedial therapy is not complete.

A TRUE STORY

Charlie Wyant

From my experience the most likely to be effective is "Noise Aversion therapy". The academics think they discovered it, but in fact it is a dog trainer's cure for problems handed down from one to another. Who first used noise deliberately to change behaviour I cannot be sure. I was taught by Charlie Wyant who, in the early sixties, would tell the following tale during his training courses.

A lady, who had arrived in an expensive car, knocked on his door and asked if he was the expert with German Shepherd dogs she had heard about. He was suitably modest but asked her what her problem was. She said every time the telephone rang her dog would bark incessantly and not stop until her telephone conversation finished. She loved her big male dog but, living on her own, could not go on like this. Charlie said "well that is simple, when the phone rings shut him in the other room". Her response was that she could not catch him!

Barking at the phone

No control

Charlie immediately realised she had no control over the dog and said he could help her but she would have to leave the dog with him for three weeks. He assured her he would be able to solve the problem and he told her what the fee would be to kennel and train the dog. For most people this would be a considerable sum of money but this lady's financial resources were greater than that of most people.

After she had gone he rescued a rusty tin tray from the garden and then cleaned an empty dog food can. He made a string handle for the can and put a handful of stones in the bottom. Having secured the lid again he fixed a hook to the ceiling in the room that contained the telephone.

Beneath the telephone he placed a coffee table with the tin tray on top. Securing more string to the can's handle he passed it over the ceiling hook and out over the top of the rooms door. Then he rang his dog training friend and asked him to phone at predefined short intervals leaving the phone calling for at least 10 rings.

Circumstantial

Following this set up, and remember I use these words deliberately, he placed the dog in the room with the can now hauled up to the ceiling and pulled the door to. When the telephone rang the dog immediately barked loudly and incessantly so he let the string drop so that the can crashed onto the tin tray. This sudden noise brought about silence.

After a while he slowly and quietly hauled the can back to the ceiling again and waited for the next call. When the call came the same things occurred in sequence and silence was again the result. With the can once more in its starting position, when the third call started ringing there was absolute silence from the dog.

Still holding the can to the ceiling he gingerly peered around the door to see the dog staring at the can seemingly stuck to the ceiling while the telephone continued to ring. He smiled and went back into the room, praised the dog and picked up the telephone. He said "don't bother to ring any more, I've fixed the problem". During the three weeks he occasionally checked that the cure was still working, which it was. When the lady came for her dog she paid her fee and took the dog away together with a now empty can. She was told she must leave this on a shelf not too far from the phone but not next to it. She did not enquire too much why she should do this.

Magic in the can

Some six months later Charlie was shopping with his wife in Canterbury when they bumped into this lady. After the usual greetings he asked her about the dog and she said she couldn't thank him enough for what he had achieved but she was mystified about the can. "Why is that my dear" he responded. "Well", she said "every time the telephone rings he just stares at the can and I cannot understand what magic you have put in it". Charlie gave a knowing smile but no explanation. Now that is an excellent tale and a first class example of noise aversion therapy becoming sight aversion therapy. I never forgot it and during the very first behaviour consultation I carried out it was a major part of the answer to the problem. However, the can was held in my hand so that I could control the shake and degree of volume and it was also ready for immediate re-use. Charlie would turn in his grave if he knew his cure now had an academic label.

WHAT IS NOISE AVERSION THERAPY

Noise aversion therapy

Noise aversion therapy can be any noise that the dog will be averse to. High pitched sounds have been used successfully, as banging a piece of wood against something. Ultra sonic sound is advertised in the press as working for all dogs but I can assure readers that while it will work for some there are just as many dogs who take little heed of such noise. Rape alarms and ships fog horns have been used and all come under the same noise label. I have known those that have banged the old metal dust bin lid to good effect. For dogs that dislike vacuum cleaners its noise could be termed a noise aversion therapy, in fact I once successfully used an owners vacuum for just that purpose.

Ascertaining levels

I always try and gauge a dog's likely reaction to noise by asking the owner about the dog's reaction to thunder and fireworks and loud bangs, or a dropped metal kitchen utensil on a hard floor. The answer is not infallible unless they indicate extreme fear. Then one must decide if noise is to be used or not. If yes, then inevitably it must be at a level that is almost quiet.

If having been called in to deal with the problem of a dog frightened of fireworks and the owner also says the dog has an aggression problem, then using noise to stop the aggression is counter productive to teaching the dog that he need not fear noise. That is an extreme example but valid nonetheless. How much or how little noise is correct for the dog in question is always crucial, and understanding sound sensitivity levels will teach this. Refer back to Chapter 6 for more information on this subject.

Rattle can

It will be obvious that my preferred type of noise is a rattle can. For average size or larger dogs a normal size soup can or slightly larger size with eight to ten pebbles in it the size of a thumb tip is usually suitable. However stronger tin than that used for those products will mean the tin is less likely to split. In this respect cigar tins such as those used in bars are ideal. For smaller dogs, or those with higher levels of sound sensitivity, then a smaller can with some path gravel inside will suffice. It will have about a third of the volume of space of the larger can. Although often with a plastic lid these days, I have found small mustard tins suit the purpose.

My preferred method of use is not to throw the can for I want it immediately available for re-use, and I will want to maintain its silence when the dog is not misbehaving. Therefore I will always hold the tin upright so that having shaken it quickly just once or twice the pebbles remain on the bottom. Think of the sound made as the word "No" which is just another sound to the dog. However the sound equivalent to "No" which is delivered by the can is up to a hundred more times powerful in its effect than the word that is delivered from the mouth.

I know some people use plastic water bottles with stones but they make less noise and may not have much effect on the least sound sensitive dogs. As I have said, there is no one cure all method of dog training and that is the same with behaviour modification. In my experience a rattle can has a high level of success in varying degrees with about 80% of dogs. The other 20% can be placed in the bomb proof category, with extremely loud noise right next to them having no effect.

Noise aversion therapy

Diminished by distance

We all know that the sudden effect of noise is diminished by distance, therefore to be effective noise needs to be used close to the dog. Part of the skilled use of it is to prove to the dog that the person with the can is all powerful. If the can is shaken from a distance of say four or five yards at a dog with average sound sensitivity levels then it may not be effective.

That will not be the only problem, for now the dog is being taught that the noise is not a problem. Therefore the dog is being de-sensitised to that noise, which is not the idea. What is also interesting is that I have never known a dog to become frightened of the person applying the sound with the can, and when successfully used the dog is usually in awe of the can itself.

Sound volumes

That leads us to the question of volumes of sound. Unless the dog is extremely sound sensitive I set out my stall by using the can on the first two or three occasions with maximum volume. This lets the dog know how powerful it is. Thereafter experiments can be made to ascertain the lowest level of sound that can be used which still maintains the maximum effect. A skilled practitioner should be able to quickly deliver just the right level of sound. With some, rapidly producing the can silently becomes effective. In such cases it has then become a sight aversion which triggers the desired response from the dog. When that occurs it is very effective.

Remember the story of the dog staring at the can on the ceiling or the mantle shelf. He was averse to the sight of the can because he remembered its previous noisy effect every time the telephone rang. If after initial use the dog has been sensitised to the noise of the can and therefore its production is sufficient for the required response, then it also follows we must be careful with our voice and actions. It should stand to reason that we cannot hold the can in sight of the dog with one hand and praise or fondle him with the other. To do so would be the confusion of two opposing messages. Therefore when the can should be silent or non-threatening it should also be out of the dog's sight. This is achieved by keeping it on the side of the body which is blind to the dog. If the dog is moving around, the blind side will change.

Commands and the can

Most usually I do not use a command with the delivery of the noise. For the reason of its likeness to the word "No" I have not found it to be necessary. What we must remember is that if we have constantly and ineffectively used the word "No" in an effort to stop the dog doing something then its use has lost its power. To then add a powerless word to the noise of the can is again risking confusion in the dog's mind.

Sight aversion therapy

If it is really felt there is a need for a command to go with the sound then chose another word that the dog does not associate with disobedience.

Praise and reward

Successful use of a rattle can or any noise is to have carried out a threatening act. Therefore that must be balanced by a quick excuse to the dog. Normally the fact that he desists from inappropriate behaviour immediately he sees or hears the can will be the excuse, for we need to also let him know when he has got it right. All old fashioned dog training recognised that correction must soon be followed by reward, and it is no different with aversion therapies.

We must accept that to use noise is to inflict stress upon the dog. Therefore it must be the lowest level necessary to bring about a desired response. In such a way it must only be mild stress. One useful but not perfect way of knowing if the dog is stressed and its degree is to offer the dog a treat. Having previously ascertained that his normal reaction is one of greed we can gauge whether he will still eat. If his concern is far greater than mild stress he will not eat until that stress has sufficiently subsided unless exceptionally greedy.

REMOTE AVERSION THERAPY

Aversion Therapies are powerful tools when used properly and appropriately. More often than not it is necessary for the therapy to be carried out on lead for this ensures that it is administered close to the dog. However there are situations where, for whatever reason, the dog will not reproduce his unwanted behaviour on lead but will certainly do so when free. Alternatively, initial aversion therapy will probably have been carried out on the lead but, as a transition to being allowed off lead again, the dog needs to learn that what he is adverse to will still occur when the owner is not in close attendance. In such cases it is necessary for this therapy to be administered by an assistant trusted by the dog.

Where an aversion therapy has been carried out with the dog on lead, to be totally consistent with the message the dog is receiving, it will have been necessary to ensure that he could not indulge in his inappropriate behaviour without it occurring. This means it is possible that the dog will only connect this occurrence with the circumstances that apply while he is on lead with his owner close to him. It may well be necessary to prove to the dog that it can occur under any circumstances and therefore two or three specially set up different situations may be required and this will involve an assistant.

What is required is a suitable place where the assistant can hide, preferably on a bend on a path. Both approaches to the bend would be screened by trees, hedge, fence etc., so that neither dog would be aware of the other's approach. If the dog under rehabilitation training is aggressive to all dogs then any stooge dog will do. If, on the other hand, he is selective, then the right type of dog must be the stooge. It might even be possible to get an owner of a dog that has previously been attacked by the dog being retrained to help and, of course, the target dog can be on lead.

That will not be the case with dog being worked with which should be running free. As the two dogs meet, should the free dog display any undesired traits, then the assistant by being on the spot would emerge extremely quickly from his hiding place to administer the therapy and therefore prove to the dog that even when his owner is remote and he is running free, that which he is adverse to will still occur.

Imagine thinking you were safe on the common and Martians descended quickly from the sky with powerful unknown weapons then, apart from any other emotions, you would be amazed. That would equate to roughly the dog's level of intelligence with a remote set-up such as we have discussed. The dog would not be able to work out that the whole situation had been contrived and if the assistant is a family member that person's sudden appearance would be a total surprise to the dog.

Just doing this once or only in one location the dog will probably only connect it to that place and at that stage to no other. In other words they would be the only circumstances where he would expect remotely applied aversion therapy to take place. In order to implant in his mind that it can happen anywhere at least more than one alternative suitable location should be used. Also it is advisable to change the target as well.

Semi-remote for some behaviours

There are of course other behaviours where an aversion therapy might need to be used remotely or even semi-remotely. If a dog with Separation Anxiety is barking in an effort to get the owners to return to him it can be applicable to use an aversion therapy. To bring about barking cessation, the door the dog is the other side of should be suddenly opened just wide enough to apply the therapy before immediately shutting it again. That is one example of semi-remote therapy being applied. Alternatively the problem may be other than barking such as stealing food or doing something that is appropriate to being stopped by sudden noise when the owner is out of sight.

Another would be as the previous story describes about the dog barking at a ringing telephone. Yet another is by ensuring the dog triggers the noise when he commits the misdemeanour, and to do this it is necessary to set up a trap. This is particularly helpful if the dog is stealing. Such a booby trap might be several cans containing pebbles which are tied together approximately 15 centimeters apart. They should be draped around the food the dog is expected to target. Far better that it is food that is not expected to be subsequently eaten by humans. If the dog tries to reach his target the cans will come crashing down on him. After this has happened once, or maybe two or three times, it will teach him not to commit that misdemeanour, but do remember also that the circumstances where he has learned might need to be altered if stealing takes place in more than one location.

It is of course extremely important that rattle cans never accidentally make a noise as this will only serve to confuse the dog if he is not doing anything inappropriate at the time. Therefore they must be carefully moved. Also young children should not have access to them for with inappropriate shaking or just making the sound they may destroy any good work already carried out.

OTHER AVERSION THERAPIES

Rolled up newspaper

As old fashioned as it sounds, when a rolled up newspaper is used to create noise it can also be effective. Usually hitting it on a coffee table or similar can do the trick, for the sudden sound can stop some dogs from their inappropriate habit whatever it may be. In days gone by some people would hit the dog with it, but that is not necessary and I never advocate any method that includes striking a dog.

Bunch of keys

Throwing a bunch of keys to the ground or onto a nosier surface such as metal can also be effective. Once again the effect of noise is diminished by distance so, as with a rattle can, all noises the dog may be adverse to should be applied close to the dog. The only problem with throwing something is that it needs to be retrieved by the handler before it can be used again.

Water aversion therapy

This will be effective with some dogs, but in my experience the percentage of success is considerably lower than using noise. There is also a draw back that with water an aim is required while with noise it is all enveloping. Water is best squirted in a fine jet onto the face of the dog immediately he indulges in the inappropriate behaviour. To squirt the dog on any other part of the body usually has little effect, for if they are going to be sensitive about it that will only occur when water hits the facial area. While people talk about water pistols or squeezy lemon containers filled with water I prefer to use household plant sprays as they have much bigger reservoirs. (Note: any spray bottle used should be new and not one which previously contained chemicals).

Water aversion therapy

Another problem is that because water needs to be squirted onto the face of the dog, when part of his behaviour is to pull ahead of the handler application becomes much trickier. For this reason I find it a big additional difficulty when advising owners to use water who have little handling skills anyway. However, water aversion therapy does have an important place with remedial therapy as there are many uses for it. Just as with noise, there must be a balance of supporting the dog when using this aversion therapy as the dog needs to understand when he has responded in the required manner.

A silent threat

Do remember that the volume of all noise aversion equipment can be controlled and a judgement of volume level should be made. Also that silent production of any aversion equipment can, in some cases, bring forth a positive result once the dog is aware of what he is adverse to.

Checking aversion therapy

For this technique the dog needs to be on lead wearing either a plain buckle collar of leather or nylon or what is known as a half check collar. This latter device is approximately seven eighths ordinary collar and one eighth chain looped through the two ends of the collar. To be correctly adjusted when tight the two chain end rings should barely meet.

Half check collar

There are very many other devices sold for attaching to leads. They vary from what are known as head collars to various types of harnesses.

Check (choke) chains are still available, as are rope and flat continuous leads that are also part of the collar. All of these devices have their place and each individual must work out what is best for them.

However, I have aleady stated my preferences as I find them to be the most effective but, whatever the lead is to be attached to, it is a question of skilled application to ensure maximum effect without in any way being cruel to the dog.

A lead can be adjusted to be too long or too short, and part of the secret is use the right length. With the dog beside the handler, that will be about three to four inches greater length than that when it is tight. I mostly use check aversion therapy for stopping aggression to other dogs, people or children that are passing. There are other types of behaviour where it can be successfully used and they would become obvious at the time.

Suffice it to say that usually the dog will strain forward on the lead to be aggressive to something ahead even though he may have been taught to walk nicely to heel. It is no use allowing the lead to become tight if he strains or lunges forward as this technique only works by checking (snatching) the lead back when it is slack. Therefore at the first sign of the lead becoming tight the check must be applied.

If caught out by the speed at which the lead becomes tight there is a quick way of putting those few inches of slack back. In the first place the handler should not have allowed the arm to be pulled out straight as it should have remained close to the waist or groin area. To make it slack the lead should be quickly pushed towards the dog's neck, which has the effect of making it slack so that it can be immediately checked.

There is the other offshoot of this method which is similar

but used as a long line, either for a dog chasing something inappropriate or for not responding to a recall command. Having written that, I must say that I have rarely used such a method, in fact I would only need the finers on one hand to count such times and then I would have some fingers left over.

It might be necessary to check several times before the dog gets the message, but if there is little positive response from the dog then that is not the right technique for the particular behaviour the handler is trying to stop. Once again there is no one method that suits all dogs. However, I have found that one or more of noise, water or check aversion therapies works for me. For that reason I have rarely attempted any other.

Some more aversion therapies

There are other aversion therapies such as citronella or water squirting collars. There are electric shock collars and fences. I make no comment on any other device other than to say I have stated what I use and have not found any need to try other things. I have to say my natural inclination to electric shock is that there has to be a better way. However, I have witnessed a case where an electric fence was successful in stopping a dog from straying onto a very busy dual carriageway which was just a few yards from the property where he lived. If it has saved his life who am I to say that this was wrong.

Whether something is cruel or not will often depend upon its application. If owners are already successfully using a head collar or harness then I do not suggest they should change. Any aversion therapy that works without side effects and is not cruel is valid.

It could be said that for some dogs, using a very displeased voice alone is an aversion therapy. With vocal, physical or remote aversion we know that we must balance our aggression. When we use any, particularly the voice, we must do so in two different ways. I cannot emphasise sufficiently that whenever the dog has stopped his inappropriate behaviour he must either be praised and reassured immediately or, if felt to be more appropriate, the dog should be asked to do something that is known he will comply with so that the handler can be seen to be supportive again. Thus the dog is reassured.

Switching back and forth

Learning the art of immediately switching from a very displeased voice to a pleasant one and back and forth is very important. Much of dog training is about timing of one's vocal or physical actions to the dog's behaviour. So one second the trainer can convey a very displeased attitude and, if the dog immediately complies, then the trainer must instantly switch to being supportive again.

These two attitudes can constantly interchange as long as they are timed to the appropriate actions of the dog. When being supportive, not only are we differentiating between support and previous displeasure but we are balancing any previous corrective attitude.

Nice Guy again

Again, if any correction is used it must quickly be balanced by support for an action carried out correctly. To only correct is to be seen to be unfair and, dependant upon the rank of the dog under training, will possibly have a frightening effect. Over a period of time that can result in a dog who may obey but will do so unwillingly. Therefore, if necessary, an excuse to support the dog must quickly be put in place. For example a quick successful recall over one or two yards might suffice for praise and support to immediately be given. Thus a balance to correction has been applied.

INCITEMENT TO MISBEHAVE

When a dog has a behavioural problem, I will often advise that the dog should be incited to indulge in the behaviour we want to change. That will be once the owner is aware of an aversion therapy that definitely produces a positive reaction from the dog. Incitement should be as continuous as possible in one lesson but, of course, backed up by support as and when applicable.

Not ready

My reasons for inciting the dog to misbehave is if we wait for the behaviour to manifest itself naturally, inevitably it does so when we are not ready to apply the therapy. Maybe we are talking to someone, or our hands are full, or the equipment we are to use is not to hand.

Aggressive to dogs

Then there is the person who has a problem with the dog being aggressive to others of its kind. They have usually spent much time avoiding other dogs. I want the dog to indulge in his inappropriate behaviour so I tell them we will go seeking other dogs so I can ascertain the preferred remedial therapy. That is incitement. I therefore prefer to incite the dog to get it wrong when everything is geared to applying one or more of the possible therapies, and both the owner and I are solely concentrating on getting everything right to ensure success of timing, etc. For the owner this is not just on one occasion but, preferably, on several short sessions over each day that it is necessary to continue. Often only two or three minutes need to be devoted to a session, with little and often being the appropriate phrase.

How to change . . .
DOG BEHAVIOUR

Part 2
NON-AGGRESSIVE PROBLEMS

1

DOMESTIC PROBLEMS

This Chapter deals with methods of all non aggressive problems that cannot be eradicated by normal basic training or general advice from such as a dog training club or veterinary surgeon. In the vast majority of cases they will be habitual and therefore correction or aversion therapies must be applied to change the dog's behaviour. Such corrective methods should never be used until it has been established that positive motivational methods are ineffective. These are not tools that should be used for basic dog training. However, understanding each problem also teaches how to avoid the problem occurring in the first place.

A REPEATED WARNING

Aversion therapies and any practical handling techniques contained within this book should also not be attempted with a highly dominant dog or the result will be that the perpetrator of them and the owners will probably be bitten. If the practitioner does not have sufficient skill to evaluate the likelihood of the dog biting they should seek greater expertise on this subject. So that I can easily refer to this warning I call it the rule of 84 (see Part 1, chapter 4).

None of the remedial techniques discussed in this book should be advised or attempted by anyone other than those who have a deep understanding of the nature of the problem dog and his likely reactions. Other family pets must be taken into account and often crucially the owner's position within the pack must be fully understood.

Also some of the advice in this book includes puppy behaviour, but greater details of how to successfully bring up a problem free puppy can be found in my book "How to Have a Problem Free Puppy". This title reflects the fact that with careful supervision so many problems can be avoided with the puppy taught by guidance and motivation.

LEADERSHIP AND SUPERIOR PACK STATUS

From hereon, where reference is made to the psychological part of solutions for problems, any reference to the owner(s) changing the dog's perception of his own superiority within the pack should include much of the advice about living with a dominant dog. It must be remembered that the highest form of rank is termed dominance, and dogs with lesser degrees of rank yet sufficient amounts of it to be pushy without being aggressive will still place importance upon many of the superior actions of a dominant dog. I am referring to such matters as sleeping quarters, retaining toys won in a game, having the right to eat first, pulling on the lead, seeking to gain height over the family and many other things important to a rank conscious dog. I have previously discussed rank in Part 1, but further advise on this subject can be found in Part 3, chapter 3, entitled High Ranking and Dominant dogs. It is all part of changing the dog's perception of his status within the pack.

RELEASE COMMAND

An aid to training and changing a dog's behaviour is to teach him a release command. To avoid unnecessary confusion a dog needs to understand when he no longer has to obey the last command given so that he knows he is free to do his own thing again. This I call a release command as it releases the dog from previous instructions. It may be that he has returned when called, or he has been told to sit, lie down, or stop where he is told. This may be in the park or at home. If having previously been given an instruction to do something and the handler has forgotten about him, on the one hand the dog can become confused as to whether or not he should continue to obey. That means he is likely to break the command, setting up an unwanted habit. On the other hand he may be continuing to unnecessarily obey.

Off you go

A release command can be a word or a phrase such as "Off you go" or "That'll do" or similar, for the words are not important and he will soon learn what the sound of them means. When he does, the handler is in a position to lessen any confusion in his mind about continuing to obey a previous instruction or not. It also helps him understand that he has not finished obeying until given the release command.

ATTENTION SEEKING

The problem and causes

Many dogs will indulge in this pastime with arguably only the most secure dogs not needing constant attention. It is carried out in several ways with some barking, others whining. Then there are those that will prod the owner's hands or body while others will just look or stare. It is caused by sufficient levels of insecurity within the dog and sometimes this is complicated by over average levels of rank as well. This makes the dog or puppy pushy with his attempts to get success. Lower ranking but sufficiently insecure dogs will be more gentle with their attention seeking.

Solutions

Dogs that constantly seek attention have usually been very successful in their previous attempts and therefore a habit has been formed. The following advice is to change the habit.

***Psychological*:** If any eye contact is made with the dog he will be encouraged to persist. Everything starts with eye contact, so if it is intended to teach the dog that he will not get a good result then it should be denied to him. This can be taken even further by snubbing the dog. See Part 1, chapter 4. If the dog puts his paws on people while they are sitting, or jumps up on them, then refer to the next problem about jumping up.

Practical: With most dogs, just working psychologically over a period of time will be sufficient. If the dog is really persistent then as well as putting the above remedial work in place then aversion therapies can be used as well. See Part 1, chapter 8 for details of aversion therapies and their application. Also see chapter 3 in this part which deals with barking for attention.

JUMPING UP

The Problem and causes

This behaviour, which easily becomes habitual, is almost certainly taught by people who interact with the dog when he is a puppy. Because he will be cute, small and very lovable he will be fondled by them and generally spoken to with a reassuring voice. When that occurs everything the puppy is doing is being reinforced. Therefore inevitably there will be times when the puppy only has his back paws on the ground, with the front ones on the ankles or legs of a human whether they are standing or sitting. Of course this can also happen with visitors to the house, and later on with those who want to greet him in the streets or park. Every time he is fondled and reassured with his front paws on a human the puppy is being taught that the owners are pleased with this behaviour.

Jumping up

Solutions

By far the best solution is to never let it happen in the first place, therefore there is no problem learned by habit that requires a remedy. See Part 1, page 25. Usually the annoyance factor only becomes apparent to the family when the dog is no longer a puppy and thus larger.

Psychological: Owners and family should turn away from the dog, and refuse to make eye contact. If a stranger entering the house of a dog that jumps up, initially refuses all contact, he will get a better result than those the dog persistently jumps up at. That will be because there has been no incident of circumstantial learning with the stranger. However, this type of appearing aloof only works for people with less persistent or below average ranking dogs.

Practical: If psychological methods are not sufficient to deal with the problem then other action must be taken. The dog must now be re taught that every time there is an occurrence of annoying behaviour I refer to as 'A' then 'B', the counter to that behaviour will be applied. Thus the dog learns it is no longer a good result to jump up.

Therefore as the dog is designed to be on four legs the best practical solution is to quickly take hold of his front legs as he jumps up so that he is held away from his target. Very quickly he will display his discomfort and desire to get down by pulling alternately on one of his two legs and/or play mouth one of the hands holding his leg.

After he has been uncomfortable for four or five seconds he should be allowed down and quickly incited to jump up again so that 'B' can be re-applied if he repeats his actions. This continues until the dog refuses to jump up, whereupon he is immediately praised and/or fondled. Often praise will inadvertently signal to the dog that it is OK to repeat his unwanted behaviour, but the target just continues to apply 'B' when appropriate.

There will soon come a point where he realises why he is praised and fondled and why 'B' is applied. Again his refusals to indulge in 'A' must always be rewarded. This should include the use of the hands, firstly to apply a balance to the target's high ranking behaviour and secondly to ensure the offending hands are also seen as supportive and not to be feared. This procedure is almost like an intelligence test for the dog. The number of times 'B' has to be applied before the dog understands why to some extent indicates intelligence. However, that can be confused by other facets of the dog's character.

It is important to understand that, as the dog connects circumstantial learning to where the initial teaching first takes place, it is necessary that he is taught in a variety of places and that may mean different parts of the house or wherever the problem occurs. Exactly the same technique can be used when the target is sitting in a chair, in fact teaching the dog that the same thing happens when there are a different set of circumstances such as when seated inevitably widens the circumstantial learning.

Warning

Rule of 84 applies. This technique should not be attempted with a highly dominant dog or the result will be that the target will probably be bitten.

CLIMBING ON FURNITURE

The problem and causes

Without doubt this problem is caused by the dog being allowed to find that comfortable beds are available to him. It is therefore a good result and resting on furniture becomes habit.

First the decision must be made as to whether the family consider the dog being on the furniture is acceptable. If they do that is their decision. There are some who like the dog to have his own chair and try to limit him to getting on that one only.

It is more difficult to teach that only one chair is his, and when that is the case what follows is of even greater importance than for those who do not want the dog on any furniture.

Solutions

Just as with the previous problem the best solution is to never allow the dog to do it in the first place. When that rule is applied before the dog gets on furniture for the first time then habitual learning will not take place.

Where dog owners make a big mistake is during the first year of his life with them. They allow the dog unsupervised access to rooms that contain comfortable chairs that he can climb on. Behind their backs he finds it to be a good result to be on an easy chair and so the rot sets in.

Until the puppy has learned what the rules are in various rooms he should not be allowed in them without supervision. This can apply with an older dog which is taken on. However, if in his previous home he was allowed to get away with climbing on furniture, even after a year he may regress.

Practical: Where the habit is already solid then an aversion therapy should be used, preferably from behind him so that to get away from it the logical answer is to get off the furniture. It is purely an 'A=B' principle that must be consistently applied until the new requirement is understood.

Dog on furniture

HOUSEHOLD DESTRUCTION

There are usually two reasons for this behaviour. One is that it is something many puppies indulge in and eventually grow out of. The second is what is known as Separation Anxiety which is descriptive of the anxious dog left without pack back up. When referring to pack that might be other canines but far more likely it will be the human element of the pack. If it is the latter another dog as company will not solve the problem.

There is a complication in that a decision needs to be made as to whether puppy destruction is Separation Anxiety or something the puppy will eventually grow out of. That is part of the skill of a behaviourist who will need to study the levels of insecurity of the puppy. Greater detail of this quite common behavioural problem can be found in chapter 6 under the heading of Insecurity and Separation Anxiety, therefore it is not dealt with here. This part of the book just deals with destruction that does not have insecurity as one of the causes.

The problem and causes

If destruction is randomly scattered around the areas the puppy has access to then it is less likely to be Separation Anxiety. Destruction brought about by that temporary mental problem will be more likely to be directed towards the door the puppy has seen the pack leave from. Also in such cases the puppy will chew the furnishings while an owner is present as well as when left. It may be that it is just puppy chewing which can be natural or brought about by teething.

These are not infallible guides for there will always be grey areas to consider and, certainly, some dogs suffering from Separation Anxiety will be disruptive to a whole room and not just the door the pack left from. Certainly neither dog nor puppy chew to get back at their owners for leaving them, but it is surprising how many people dream that up as a possibility.

If it is the case of an older dog that has not chewed before and then the problem starts to occur it is more likely to be Separation Anxiety. Usually, within the pack (family), there will have been a change of circumstances that can be easily pin pointed. If so that will be what sets off the problem. There is also the complication of an older dog also starting to chew, etc. while the owners are in sight. Sometimes this can be the result of circumstantial learning from habits acquired while anxious and separated from the pack.

Solutions

In the first place the dog should be confined to one area so that damaged areas are limited. When the puppy or dog is chewing habitually while in the company of one or more owners then practical solutions may be appropriate. However, the practitioner must be reasonably certain that there is no temporary mental anxiety illness.

Practical: If it is natural puppy chewing or driven by teething then eventually it will cease. Distraction should be tried, but unless the motivation of the distraction has greater value to the dog than the desire to be destructive it will not work. The next step would be to coat the areas chosen for destruction with something unpalatable to him. If successful, again it will be an 'A = B' principle. Bitter apples and bitter aloe solutions are disliked by many dogs. Some people have had success with hot spices such as Chilli Pepper or Tabasco sauce, but equally such attempts at making the target furnishings unpalatable do not work for all dogs. Experiments need to be conducted to find what solution the dog is adverse to, if any.

Aversion therapies can be used, but if it is to be noise care must be taken not to set up fear of sudden noise in a puppy that previously did not have that problem. That is common sense. Alternatively it should be a jet of water into the dog's face, and even those that like water will usually respond to this therapy.

When the dog desists from his destruction then he should be praised.

EXCITED BEHAVIOUR

The problem and causes

For our purposes we can break down this behaviour into three categories with the most intense being hyper-activity, and then in descending order, excited behaviour, followed by a busy dog. The first two categories are self descriptive while the last one describes a dog which falls short of being classified as excited but is a generally happy dog which likes to be doing something, usually at speed. These latter types of dogs make excellent workers when of a suitable breed for the proposed task.

A cause of this problem will be the levels of natural activity given to the dog as a result of breeding or of nature's interference with it. Whilst jumping up can be part of this problem I separate it from excited behaviour for not all excitable dogs jump up.

With regard to category three, a busy dog's activity only needs to be contained by general training so that when told to lie down there is a positive response. It might be that there has been an over stimulation to play, with one or more owners being seen as the dogs play mate. For these reasons, and to save such dogs from unnecessary further treatment, the following only applies to the first two categories.

Solutions

Generally it is a veterinary surgeon who will diagnose hyper activity. If this is not the case a veterinary surgeon needs to be consulted for, among other factors, it can be determined if the dog is clinically hyperactive.

Drugs and Dietary: If a veterinary surgeon has prescribed drugs, almost certainly the dog's diet will have been ascertained as well. Certainly the dog should be on a low protein food for it is this ingredient that gives much fuel for activity, hence working dogs are generally fed higher protein levels. There is also a small possibility of artificial colouring and flavouring affecting the dog's behaviour.

Practical: With the first two categories, often the behaviour has become at least 50% habitual so that even if other veterinary treatment is to be used then it may still also be necessary to change the dog's habit. It is difficult to advise exactly, and some intuition needs to be used by the practitioner to determine whether any aversion therapy should be used and if so which one.

Excited behaviour or worse can manifest itself in so many different ways. It might be excited barking which is discussed under the heading of barking later in this chapter, tearing around all the time, just non stop attention or play seeking. Certainly at the least experiments should be carried out to see what therapy is valid.

With the majority of cases I have seen, an aversion therapy has brought about success and, when that happens, it is almost confirmation of the habitual effect of the behaviour.

Psychological: I regret to say that those who advise ignoring the bad behaviour as a remedy will have overlooked the score of the driven motivation against the score of the applied motivation which, in this case, would be just to withdraw any response to the dog's activity. In the vast majority of cases the driven motivation will always be the winner. Therefore ignoring the behaviour is often non-productive.

CALLERS AND PASSERS BY

This heading is to cover the dog that is territorially over active without being aggressive.

The problem and causes

In the first place we must acknowledge that it is natural for dogs to display territorial awareness and to give warning on the approach of strangers. While owners have differing views about the amount of territorial behaviour that constitutes a nuisance, there is a generally accepted level of this activity that is above average and therefore becomes an irritant.

The worst form of this behaviour is when there is aggression at the end of it but that is dealt with in Part 3. Usually it will be callers and/or passers by that get the dog's attention and, in some cases, other dogs passing can receive the same treatment. Sometimes it is dogs that just attack the post coming through the letter box. Typically postmen themselves are targets, as are refuse collectors and utility meter readers. Often much barking and generally rushing around in a very excited manner is part of the territorial symptoms and, if the dog is running free, sometimes that behaviour can be interpreted as intimidating. For that reason owners must be aware that it is possible to fall foul of the Dangerous Dogs Act.

With a dog's perception of getting a good result from intimidating actions, stopping the behaviour also stops habitual reinforcement of a good result.

Guarding territory

That in itself is unlikely to affect a cure but it is part of the jig saw puzzle.

Solutions

Sometimes extreme driven motivation of the dog plus any habitual factor makes this problem difficult to stop. It must therefore be approached using every possible advantage, which definitely means both psychologically and practically and using common sense. The dog needs to be fully aware that there are one or more higher ranking members of the pack. Having established that, practical work can be put in place which will reinforce the dog's perception of a rank higher than his own. That makes it much more likely that any applied motivation will be greater than the driven amount (see Part 1, chapter 4).

The dog should not be allowed access to areas of the house where he can continue to reinforce his anti-social behaviour. For example, furniture backed up to a front window that thc dog has access to allows the dog to indulge in his habit. Particularly smaller dogs will jump on the furniture to be able to intimidate at the window. Being able to get at the post as it comes through the door should be stopped for the same reason. Those with houses and secure grounds where the dog is running free anywhere adjacent to areas that passers by or callers use, allows the territorial dog to continue with his hobby and cement the habitual factor.

Restricting access from all such areas does not cure the problem but, as all dog training relies on consistency from the teacher, to allow access to targets negates any other attempts to change the dog's behaviour. Inevitably the dog will still hear, and in some circumstances see, that which annoys him and therefore he will still behave in the same manner at a greater distance.

Psychological: It is extremely important that the dog understands he is not the superior pack member, particularly in these situations. To understand how a dog's perception of being a lesser ranking pack member can be achieved refer to Chapter 4 in Part 1. However, remember the circumstantial learning factor will also play a part.

Practical: The aim is to teach the dog that he does not have sufficient pack status to be the one that initiates deterrent displays to make the target go away. He has to understand that only superior pack members have this right. Some of this will have been established psychologically, but much must be practical negative motivation.

Experiments will need to be carried out to find whether it is better to check the dog on the lead each time he indulges, or if an aversion therapy would be better applied. It will quickly be apparent whether he will back down to the aversion, for if he does not do so very quickly the particular negative motivation being applied at the time is not going to work by continuing its application. If he does back down then he must be supported for getting it right.

In many cases help is required in the form of stooges to ring door bells, etc., or do whatever causes the dog to act territorially.

I cannot emphasise sufficiently that consistency is paramount, therefore to leave the dog to indulge in this hobby while the owners are out is to undo any work they may have done while they are at home. Having established the best way to proceed, regular training sessions should be set up when the dog should be incited to make mistakes.

CHASING THE FAMILY CAT

Using the word family in this title is deliberate, for when a family does not have a cat but the dog chases those he sees it is a predatory problem which is dealt with elsewhere.

The problem and causes

It is still a predatory problem but in many instances, other than the fact that the family cat is often around, there would not be much of a problem. While it can be an over average predatory instinct that causes the problem, a dog with average or less of that instinct can start to learn a bad habit. It will most likely be due to regular close proximity of the cat. The dog gets pleasure from making the cat move, therefore a habit is developed.

Solutions

Both psychological and practical work will need to be put in hand, for it is highly unlikely that distraction from the cat will outweigh the dog's driven motivation. However it should be tried first if the problem is only mild.

Psychological: The dog must understand that one or more owners have the ranking order right to demand that the dog does not have the pack superiority to decide to play this game. Therefore chapter 4 in Part 1 applies.

Practical: If the dog cannot be stopped by motivation or by voice alone, then aversion therapies should be tried applying them in front of the dog. If it is to be noise then the effect of noise upon the cat must be taken into account to ascertain whether or not it should be used. If noise has become a sight deterrent then silent production of the noise implement may well suffice. It is also possible that the dog may respond by checking with the lead. Whatever the dog responds to in a positive manner, from then on consistency will be the key.

Chasing a cat

SEXUAL MOUNTING

The problem and causes

Those persistent with this behaviour will most likely have an over average sexual instinct. Others will have far less of this instinct in varying amounts. Sometimes it is part of high ranking or dominant behaviour.

This is a behaviour that is occasionally indulged in by puppies and then they grow out of it. With others it is more pronounced and a minority can be working habitually by the time they are adults. The targets for mounting will be the owners or visitors, and often it is their legs. Sometimes the dog has a favourite possession that gets the treatment, such as his bed.

Solutions

When a puppy has not grown out of it then a veterinary surgeon should be consulted about castration. This may completely deal with the problem. On the other hand, if it has been regularly occurring for some time then there may be residual habitual occurrences after castration. In a few cases bitches will sometimes carry out the same actions as males, but I have never known this to become a major problem. When it does it is usually a hormonal imbalance and veterinary advice should be sought. If it is part of dominant behaviour then please read part three on that subject.

Practical: Distraction may be sufficient for a low driven motivation. Escalating from there, severe use of the voice may be sufficient but not if the dog continues after the admonishment has been delivered. That will be to teach disobedience and diminishing power on the part of the owners. If necessary use an aversion therapy appropriately, and thereafter ensure consistency of application.

FEAR OF STRANGERS, DOGS AND SITUATIONS

This is a very common behaviour and in this part of the book I refer to a dog that avoids or flinches away from anything or displays fear by general body language. A dog that has escalated to fear aggression is dealt with in Part 3, chapter 4 under the heading of fear aggression. I include under this latter title all threatening levels from barking, lunging towards or pursuing a target and those that will bite as well. This part of the book deals only with dogs that might be better described as wary, shy or mildly fearful.

The problem and causes

Fear displayed towards babies, children, adults, visitors, traffic, sounds, dogs and other animals are all included here. As stated in Part 1, I believe that any dog that displays fear must have been born with sufficient latent levels of fear for it to become active subsequently or very early in the puppy's life. Often there are further complications of rank and insecurity. More often than not a fearful dog is of below average rank and has over average levels of insecurity. But as we know, nothing to do with behaviour is definite so there can be exceptions to that rule.

Sometimes poorly thought out breeding can be responsible for producing puppies with fear that is apparent before they leave their mother. With others a bad experience will bring out any latent fear, as well as the mother inadvertently teaching the puppies that strangers are a problem. I also believe that nature will sometimes interfere for she requires a proportion of fearful dogs to make up the spectrum of different characteristics within a litter.

A lack of socialisation and/or withdrawing the dog from that which it is fearful of will cause latent fear to become active or exacerbate an existing problem. Therefore, while a natural human response is to avoid problems, doing so with a fearful dog only makes matters worse.

Solutions

Psychological: Ensuring that we do not transmit or reinforce fear is very important, particularly if we are definitely seen as the dog's superior both territorially and non-territorially.

Fear can be displayed towards a new family baby. Of course safety of the baby must always be paramount, but not to the extent that the cause of any such problem towards the baby is that the mother always withdraws the baby from the dog. If that occurs it risks the dog associating approaches from the baby as a problem. That can then escalate to the extent that, when the baby eventually crawls towards the dog he will see that as a problem and take evasive or defensive action.

Whatever the dog is afraid of, controlled socialisation is the only final cure. If necessary I would suggest that you ensure you have sufficient knowledge about fear and controlled socialisation by referring to chapters 6 and 7 in Part 1. While it might be deemed appropriate to use a flooding technique with any other measures being advised, it is always wise to consider whether a gradual approach to that which the dog is fearful of is more appropriate. That may sound like conflicting advice, but sometimes opposite approaches are valid with remedial work. It is for a skilled practitioner to make such a decision.

With heavy traffic or plant and machinery, a gradual approach over a period of time will almost certainly be the best remedy. Fear of people might also include a gradual approach of the dog being flooded with people who ignore him. Again that might seem like opposing advice, but on one hand he needs to find that non interacting strangers pose no threat. On the other hand, those that approach him to eventually interact should be controlled by the practitioner or handler so that the whole process is deemed by the dog to be non-threatening.

Practitioners need to understand that often they are strangers themselves. Therefore until the dog has accepted eye, hand and voice contact from them without displaying any sign of apprehension, they must also follow the golden rule of avoiding all those three things before eventually gradually introducing them.

Practical: While the final solution to fear based problems is controlled socialisation, sometimes there are other things that need to be done before socialisation can commence. For instance, it is impossible to socialise a dog intent on attacking that which it is fearful of until the aggression has been eradicated. If the dog's fear includes some intimidation of his target, mild application of an applicable aversion therapy may be the answer before socialisation can commence. All aggression problems are dealt with in Part 3, and that caused by fear in chapter 4.

SUCKING BED OR COMFORTER

The problem and causes

Usually this is puppy behaviour and sometimes it goes into adulthood. Like a baby having been given a dummy to suck on is allowed to keep it too long until it becomes almost addicted to it. It is not a common problem but there are odd instances of dogs that spend much of their waking time sucking on a comforter. I have encountered it on several occasions. This behaviour is not to be confused with the dog who has toys in his mouth with the intention of playing.

Sucking a comforter

Solutions

With this problem neither psychological nor practical solutions are required, for it is simply a matter of using common sense by not allowing the dog to have his comforter. If that includes his bed, a non cloth one should be used until the habit is broken.

ROOM OR AREA RESTRICTION

The problem

Some owners like to restrict the dog from access to certain rooms or parts of the house.

Solutions

As with quite a few inappropriate behaviours, this is one that is best cured by never allowing the dog to cross the forbidden threshold. That certainly means absolute consistency, for once the dog has crossed the demarcation line it is much more difficult to teach him not to do it. Ensuring the dog does not cross the threshold once any barrier is removed might be achieved with aversion therapies. That would be circumstantial learning and may not have the required effect when the dog is left alone for several hours, for the circumstance missing would be the person that applied the therapy. Certainly the lower the dog's natural rank the easier it will be to achieve, but it might prove impossible with the higher rankers.

Practical: The most usual family requirement is to restrict the dog from the upstairs part of the house and consistency can be achieved by using a baby stair gate providing it is always kept shut. How long that needs to be in place will differ from dog to dog, but at some time a judgement has to be made as to whether the time is right to remove it.

There should be a transitional period where sometimes the gate is there and sometimes when it is not. In this latter case the handler or practitioner should always be able to observe any attempts by the dog to cross the threshold. If he does an aversion therapy should be applied immediately with the most appropriate one being noise. This could be administered from a distance if a situation is developed such as described in Part 1, Chapter 8.

PLAY MOUTHING

Often those dealing with behaviour problems will be asked to advise on this problem, or when the practitioner visits the house of a family with a dog to advise on another behavioural matter finds that play mouthing is an additional problem.

The problem and causes

Play mouthing is what puppies do with their litter brothers and sisters. When they leave that litter and go to a new home we should encourage them to play with us, but not in a manner which triggers play mouthing. It must be frustrating for puppies when they realise that we can manipulate them with our hands, which they cannot do. For them the mouth is the only option. Natural born rank will often play a part in the degree with which play mouthing varies from one to another puppy,and sometimes it is accompanied by play growling.

Solutions

Habits are being formed and so right from the beginning the puppy should be taught that this behaviour is unacceptable, and even more so if accompanied by growling.

Psychological: A puppy does not intend to hurt when they play mouth and they need to know how hard they can do so without hurting. Often just saying "Oow" in a loud high pitched and hurt manner will teach him he has bitten too hard even if that is not the case. Not every puppy will respond to this but many will, so it is worth a try. Also it will be found that some who do not respond to the "I'm hurt" shout at a younger age do so when they are older and in particular when superior rank has been established. If the puppy is older then the superior pack status should have been put in place and then cessation to any practical measures is more likely to quickly succeed.

Practical: If necessary a growling "Argh" should be delivered in a manner consistent with his rank particularly close to the puppy for, if he has already learned that he is not the superior animal, this may suffice. Distraction of treat or toy might stop it, but care should be taken that he does not learn to play bite to get such distractional motivation. If this is not sufficient then it is necessary to accompany that command with a tap across the bridge of the nose with a finger for, to get the message across, it may be necessary to make his action uncomfortable as well. The weight of such a tap should be commensurate with his fragility of age and size and also rank.

If this has been habitual for sometime and the puppy is now better described as an adult then a mild aversion therapy can be tried. Whatever method is found suitable, immediately the puppy desists he should be praised, fondled and reassured that all is well but should he recommence play mouthing, immediately return to the method that achieved success.

Rather than wait for the occasions when the puppy play bites it is also preferable to set up training sessions. Such sessions should incite the puppy to indulge in play mouthing so that he can be taught that his behaviour is not acceptable to other pack members. If family children are of an age whereby they are either too young or not yet able to carry out such training with good timing, then one of the adults should do so on their behalf.

AVOIDING GROOMING, TOWELLING, etc.

The problem and causes

This is often a problem with puppies that have not learned their owner's requirements about the cleanliness of their body. However, an inappropriate behaviour not rectified when a puppy, can become an adult habit. Mostly it is avoidance and trying to play with comb, brush or towel.

Solutions

Usually all that is necessary is the owner's insistence that this ritual is going to be completed and gradually matters improve. What must not occur is that the dog learns that he can stop the owner doing this to him, so it is basically a question of whose will prevails.

Psychological: It is important that the dog acknowledges the superior pack right of the owner to interfere with his body. (See Part 1, chapter 4).

Avoiding grooming

Practical: Sometimes the puppy or dog will also try play mouthing when trying for a cessation, and that should also be stopped. If the dog is highly or dominant ranked then they might start growling at the perpetrator of the perceived indignity. However, if it is possible, get the dog to back down vocally or just continue grooming throughout his grumbling.

While no risk should be taken, it is unwise for the dog to learn that whatever tactics he uses that he can make the owner desist. If there is definite aggression then a practitioner should advise whether an aversion therapy is suitable and which one should be used.

STEALING

There are dogs that have learned to steal many items. Often it is food to eat but it may also be other items such as clothing from the washing basket. Some will be eaten, and that will almost certainly include stolen food.

The problem and causes

Arguably the most annoying is the dog going into the kitchen and stealing food from a work top. This might be just a bun or far worse the family dinner. There are also dogs which have learned how to open the refrigerator door and then consume much of the contents. Of course when a dog eats a whole pack of butter, or similar food in such quantity it will inevitably upset his stomach, the result is often diarrhoea and therefore one problem causes another. This behaviour usually starts at puppy-hood and often is the result of inadequate initial supervision of the puppy as he is allowed to roam around the house. What starts as a rewarding discovery for the dog triggers a good result and then subsequently habit kicks in.

Solutions

This problem is a classic case of prevention being far better than cure. Owners of young dogs must study their kitchens and effectively make a risk assessment. If it looks possible that the fridge door can be opened then something must be done that can prevent such an occurrence.

Psychological: A new adult dog or puppy introduced to the house carefully supervised in every room he is allowed to enter cannot get much wrong. While he is learning what he can and cannot do, a little ingenuity can ensure that he does not steal from the washing basket, bedrooms or kitchen. Ensuring that access is denied to all tempting items or certain rooms is usually all that is necessary. It is not a lot different to teaching a dog not to get on the furniture by always initially being with him to ensure it does not happen. Inconvenient? Yes. Effective? Yes.

Practical: For the dog that has already developed such an unwanted habit, the most obvious solution will be catching him in the act and remonstrating with him in a manner commensurate with his rank and age. That might well be a reasonable gentle telling off or something of a firmer nature that is not in any way inappropriate or cruel.

Aversion therapies can be used but often with the food stealing dog he will have already learned the family's clumsy attempts to unsuccessfully stop him. This will have resulted in him associating their disapproval with what he is doing but the driven motivation will usually be greater than his willingness to desist. This means that a more isolated correction will be needed. One way of doing this is to use several rattle cans tied together with string so that they can be draped around the item he is likely to steal. This will probably be more easily achieved with stealing food from a kitchen work top. Therefore deliberately set the trap for the offending dog ensuring that any attempt to steal should result in the cans crashing around him. Remote aversion therapy in Part 1, chapter 8 gives further detail.

2

INGESTING, SLEEPING AND GARDEN

In this chapter we will look at behavioural problems connected to eating, drinking, sleeping quarters and the garden.

EATING AND FEEDING

While it may seem obvious that it is a good idea to have a definite place where the dog is fed, this may not be so when he is given a biscuit, chew or bone. If you always ensure that he does not get the tasty item until he is in the right place he will always automatically go to that place once he knows he is to be given something.

The problems and causes

Some dogs only eat to survive and easily become fussy eaters. Often this means that the owners become so worried that they try all kinds of food in an effort to find something the dog appears to enjoy. With some they can finish up on quite expensive human food such as fillet steak. Other dogs can be protective of their food and this can occur in varying degrees. Sometimes it is only bones or chews that make the dog protective. How intense or otherwise the dog is to protect his food will depend on his natural rank.

Normally the higher the rank the greater the problem and therefore this would be the cause. However, it is not always that straightforward, for a dog that otherwise might be quite low ranking but has an extreme driven motivation for food, can display aggression over perceived attempts to take his food away. Therefore, on this factor alone, an experienced practitioner on behaviour would be best to judge whether it is the dog's rank or just extreme greed which is causing the problem.

If it is rank then it might be high rank or at the worse dominance. If it is dominance then only those with experience of such dogs should take a decision about how to proceed. See chapter 4 of Part 1 for details of rank, and Part 3, chapter 3 for greater information under the heading of dominant dogs.

Solutions

Psychological: Fussy eaters should never have their food left down so that they can take a little at a time. It is always far better to have one or two times a day when the dog is fed and those times should only be approximate. This is to avoid the greedier dog making a fuss because his normal feeding time has been passed. No treats or extras should be given in between, it being the objective to make the dog hungry for his food when it is placed before him.

Of course the dog should have access to water, but if he does not eat for two days do not worry for 48 hours without food does not mean dying of starvation. Usually the survival instinct will force the dog to eat what is offered, but if refusal to eat continues to a third day then a veterinary surgeon should be consulted.

Practical: Those protective of their food would have been better dealt with when puppies for it is all part of practical methods of ascertaining superior rank at an early stage. In this case it teaches the puppy that a superior pack member does have right to be close to his food and indeed to take it away, albeit temporarily. I have always believed in teaching the dog my rights as a superior but also teaching him that good behaviour is rewarded by the food being returned.

If that means growling back and, if necessary, giving the puppy a little shake to get him to back down, that is what should happen. As soon as he has backed down the food can be returned. Then let him have a few mouthfuls before removing the food again. If his reaction is the same then, likewise, the response should be the same.

Providing each time that the dog is corrected he backs down then this procedure continues until he shows no reaction when a hand is placed on his food dish, bone or chew.

If it is an older dog, perhaps a rescue with a food protection problem, then the rule of 84 of rank must be taken into consideration. That is if accepting his confrontation will he be likely to retaliate and possibly escalate aggression?

Aggression over food

If that is the case then a food exchange programme should be put in place. John Rogerson details this adequately in his book entitled "Understanding Your Dog". He suggests that the dog is first fed a less appetising food and then as this is taken away the dog is offered something much more appetising. However as advised above, if there is any doubt about whether or not the dog will escalate his aggression when attempting food exchange a behaviour practitioner should be consulted.

If that does not solve the problem then change the dog's feeding area. This would be in an attempt to remove the habitual aggression factor whereby, in a new feeding place, the dog cannot fall back on circumstantial learning. In other words the circumstances would now be different. These days raised holders for food dishes can be bought and one of these is ideal to stop the dog standing over his dish. Allow the dog to eat a few mouthfuls from his dish before offering him a dish containing a little of something much more appetising, and as he takes food from the new dish remove the other one. If only a little is used at a time then several changes of dish can take place. For extra courage to act in a positive manner I would suggest gardening gauntlets are used and the face is kept clear of the dog's aggression equipment.

If this is not successful or deemed to be too dangerous then the only other option is to avoid the problem altogether by feeding the dog in another room. A garage or suitable room would be an ideal place. With the dog shut out of the kitchen and the route to that room, prepare his food and place it where he is to be fed. Then take the dog to the room and allow him in before shutting the door. Subsequently, and only when appropriate, allow him out. This does not cure the problem but it totally avoids his continual reinforcement of superiority and danger, particularly to children.

SCAVENGING

The problems and causes

It is not uncommon for dogs to pick up anything that remotely tastes like food. Whether it is food itself or the wrapper that tastes of it. Some breeds are more prone to this than others, and those whose instincts have been honed to look for prey over many generations can easily become scavengers.

There is also the manner in which the puppy was weaned. Did he have to fight for survival in a communal feeding bowl therefore learning to quickly ingest any food item before others get it? Thus it is easy for a habit to be formed. That will particularly occur as a result of a good experience when enjoying eating what has been found.

Solutions

Of course there are many items that the dog can find on a walk or in the garden that can be harmful to him. An owner should accept that the puppy or dog may try to ingest anything and therefore need to keep a close watch to ensure that nothing dangerous is available.

Psychological: Once again a programme of establishing true leadership over the dog is an important part of being able to stop him picking up and ingesting things that he finds. In other words the aura of the owner must be that only they have the right to what is considered by the dog to be edible items.

Practical: Using treats as a distraction may be sufficient to break the habit or at least give the owner a chance to take other action. However, there is nothing like obedience to a command "Leave". It can mean differing things according to the circumstances. For instance it can mean leave that other dog or person you are annoying, and it can also mean do not pick up or drop whatever you are attempting to eat or are just holding in the mouth. Competition dog trainers need not worry that this means teaching the dog to drop a retrieve item for that will occur under different circumstances. For instance if the handler always has a definite body language set up for a retrieve then there will be no confusion when on other occasions the dog is asked to drop something. I refer again to circumstantial learning in Part 1, chapter 2

While this book is not about training exercises it must be pointed out that to teach a dog to leave something he has in his mouth he also needs to be taught to hold it as well. In other words it is basic retrieve training that develops this very useful command. Another two books in this series detail retrieve training and therefore teaching the dog to leave. These are entitled "How to Train Your Dog - Part 2" and "How to Train a Working Trial Dog".

Common sense: In the garden all likely things that may be ingested must be removed. That might mean a daily inspection before he is allowed out unsupervised. Certainly walking him in areas that are kept clean of rubbish would be helpful.

If the problem is really bad then the dog must be kept on a lead until a solution has been found. That might have to be aversion therapies where the dog is incited to pick up an offending article and then the therapy is applied using the 'A = B' principle which is explained in Part 1, chapter 8.

PICA BEHAVIOUR

The problems and causes

A very similar problem to the one above is what is known as Pica behaviour. This is when a dog picks up non food items, maybe chews them but definitely ingests them. First of all a veterinary surgeon's advice should be obtained to ensure that there no medical conditions causing the problem and therefore it is concluded that the problem must be behavioural. The most likely behavioural cause is by habit, for a dog uses his mouth much like we use our hands. It is a method of exploration. That can initially lead to chewing and then accidentally swallowing an offending article. Having enjoyed the experience it might start to become habitual so that items are ingested willingly. It will most likely take place outdoors, although some puppies and dogs do get a passion for stealing tissues from waste baskets, etc.

Solutions

Much of what I have written before about scavenging dogs holds good for this similar behaviour.

Psychological: Additional to the above, owners must ensure that when trying to get something away from the dog they do not inadvertently make it a good result for him. For instance allowing him to avoid capture or chasing him would be regarded as fun and therefore as far as the dog is concerned the whole behaviour becomes worth repeating.

Practical: Inciting a dog that has this problem to pick up and ingest something you have already suitably doctored will work for some dogs. The idea is to make it unpalatable, and some substances that can be experimented with are bitter apples, bitter aloes, chilli powder, Tabasco sauce, citronella, etc.

Distraction with food can again be tried, but its success will depend on the applied motivational score of the food offered to that of the driven motivational score of the item the dog wants to eat. Any suitable dog training command that the dog instantly responds to can be used, but some dogs will quickly ingest and then obey. Therefore the only suitable command is a solid "Leave".

Finally, aversion therapies can be experimented with, but practitioners must remember that noise effectiveness is reduced by distance and often it may not be possible to get close enough to the dog at the right time. Therefore there might be an argument for rigging up a remote application of noise so that it can be applied from a distance yet beside the offending article and dog. See more details of remote noise aversion therapy in Part 1, chapter 8.

DRINKING

Other than medical problems there are not many problems associated with drinking.

The problems and causes

A very hot and thirsty dog may attempt to drink any liquid that he finds. I have known of a case where a dog belonging to a commercial tomato grower regularly drank from the plant feed tank without the owner's knowledge. This eventually killed him.

More often problems with water are those associated with puppies. They can paddle in their drinking bowl, tip it over or generally treat it as something to play with.

Solutions

Practical: Care must be taken that possible noxious liquids and those that might be poisonous are not available to a dog running around in a garden or business premises.

With puppies playing with their water bowl a good solution is usually to get the bowl above ground, and therefore paw height, by using a bowl stand that is usually available from pet stores.

The use of a bowl stand

SLEEPING QUARTERS

In general

Where the dog sleeps both during the day and night can have behavioural implications. Certainly a dog should be given a bed of his own and taught to go to it on command.

Of course the bed should be comfortable, with one possible advantage of that being he may be less inclined to want to use the places of comfort that family members use. We all know that where a dog regularly lies soon starts to smell doggy, so to a certain extent the dog's own bed confines such smells to that area. It should be placed in a convenient and sensible place, and there is no problem if he has additional beds in other rooms.

When a dog knows where his bed is and a command that he associates with going to it, this can then be used as a place to send him to for many reasons. One such reason would be when he is generally getting in the way.

However, it should not be treated as a place of punishment but rather a place where he is secure and cannot get into trouble.

He should be taught to go to his bed by encouraging him in to it and as he does so the chosen command is added such as "in your bed" or just "bed". It can also be taught as a game by playing with him using his favourite toy and then placing it in the bed. It follows that he should also sleep in his bed at night.

The problems and causes

Allowing the dog to sleep in the bedroom can be a problem for dogs with over average insecurity problems which can bring about separation anxiety. This is a temporary mental illness that becomes apparent each time the dog is left on his own. At first glance that statement might seem to be confusing but, basically, excessive company for a dog with such insecurity levels allows him to become dependent upon it and thus, eventually, when left the problem starts to occur. See chapter 6 for more detailed information.

Then there is the high ranking or dominant dog who may who given such a privilege, will see it as reinforcement of his already perceived high rank and develop some anti-social problems in the bedroom. See Part 3, chapter 3 for further details.

If the dog is already in the bedroom at night, in the majority of cases it has occurred because in the past the dog has been noisy or destructive when left in a room away from the humans. It would have been preferable for the owners to persevere with his night time isolation rather than take the least line of resistance by allowing him to one or more human bedrooms where he then becomes quiet.

Solutions

Before the dog is left for the night owners must ensure that he has eliminated waste from both parts of his body before retiring to bed. Secondly that the dog should not have had a good result by one or more family members going to see him when he creates a fuss. That would be either to remonstrate or to console him. Both would have had the effect of giving him what he wanted which would have been to be reunited with the pack. Once a dog learns by habit how to get one or more members of the pack to return to him he will continue with his successful behaviour.

Psychological: By the dog not getting successful re-unification with his owner(s) he is far more likely to eventually give up his quest for them to return to him.

Denying a rank conscious dog from a human bedroom sends a small but necessary message that he does not have equal status at night. With both possible problems it also teaches the dog to live with his own company, and that helps when he is left alone at times during the day.

Practical: If the dog is persistent then the owner can try going quietly to the door he is secured behind and banging sharply on it as he makes any noise such as destruction or barking. If successful then no praise or support should be given for remaining quiet. It is in effect an exception to the rule of praise or support being given to a dog for getting it right. In this case it would be counter productive to the problem.

BARKING IN THE GARDEN

See the next chapter for details of this problem.

ESCAPING FROM THE GARDEN

It will be a mater of common sense that the garden is securely fenced before introducing a puppy or dog to it. However, what many will not know is that while a low fence might suffice to keep some dogs in, others will quickly learn to jump a much higher fence or gate, particularly if they are rigid.

The problem and causes

For a dog to want to escape there has to be greater driven motivation beyond the garden. That can take several forms. It might be that he just wants to return to the wide open spaces that he is normally exercised in. It may be that he knows where there is a supply of something to eat, and it might be that on a walk he came across the scent of a bitch in season and he is determined to follow it up. Of course there could be other reasons as well.

Solutions

It is another of these problems that are best stopped by never allowing them to happen in the first place. That means the dog being supervised by a responsible family member on every occasion he is allowed into the garden until it is ascertained that he has no desire to escape or wander. If at any time he does attempt it then he must be deterred on the 'A = B' principle which is explained in Part 1, page 96.

Escaping from the garden

If it is an adult dog with an entrenched habit, such as a rescue dog, then perhaps an aversion therapy might be suitable and a behavioural practitioner should make that judgement.

It is also interesting to note the well known fact that dogs taught to jump in competition rarely attempt to do so without first being given an instruction to that effect. Dogs also dislike attempting to walk or climb on anything that feels unstable. Therefore a nice solid fence or gate may be easy for the dog to scale by using his front paws to gain purchase on the solid top before hauling himself over. The top part should be made wobbly, and ingenuity should be used to find a suitable but not dangerous way of doing this.

There are also dogs that will try and tunnel their way under a fence and, if this is the case, then something far more solid than earth should be used below the fence so that the dog cannot dig past or under it.

DESTROYING THE GARDEN

I once heard it said at a lecture about growing Fuchsia flowers that gardening and dogs do not go together. More correctly the lecturer should have said untrained dogs and gardening do not go together. As a dog trainer and grower of Fuchsias I should know.

This is definitely one of those problems that are best stopped by never allowing the habit to start in the first place. That means the dog should be accompanied in the garden for all his visits until such time as the owners are sure undesirable habits are not going to occur when they are out of sight. Sometimes a dog initially digging in the garden is seen as funny or cute but usually as the habit sets in it first becomes an inconvenience and subsequently a problem.

The problem and causes

Dogs can wreck the garden in various ways from digging holes to tearing out plants and generally trampling over flower beds, etc. to relieving themselves in inappropriate places. Causes can be varied, from hyperactivity to a strong instinct to play or dig. Others would be inquisitiveness, boredom, chasing birds and animals, realising it gains the owners attention, or simply turning the behaviour into an habitual game. Just being an exuberant puppy is another, and a bitch puppy yellowing grass where she urinates can be irritating if the owners have a nice lawn.

Solutions

From the first time the dog goes into the garden and continuing until it is established that the dog understands the owner's garden rules he should be supervised. This, if necessary should be on lead or a long line, for nothing can exacerbate a problem more than the owners trying to catch the dog or puppy while he plays exactly that game.

Psychological: One of the first things is that pack superiority should have been established so that the dog understands that when an instruction is given he is expected to obey it. Chapter 4 of Part 1 gives details of how this is achieved.

Practical: A good recall should have been put in place in the house and it should also be practised on the lead in the garden on several occasions prior to being attempted off lead. In such a way the owners will have a better chance of gaining control before the dog is tempted into indulging in wreaking activities. Of course a recall to an already taught signal that there may be food in the hand is of great assistance. Details of teaching obedience to a recall command can be found in chapter 4.

Distraction with food or toy can be attempted but it will not be successful for all dogs. Certainly average to below average dogs may suitably respond. Alternatively every time the dog does something inappropriate in the garden other than eliminating, waste from the body then an 'A = B' principle should be used. See Part 1, page 96. It may be that vocal admonishment, followed if successful by support, will be sufficient. For the worse cases of entrenched habitual activity then aversion therapies should be tried. Sometimes a combination of many differing methods is the answer.

Where it is a case of elimination problems, first of all the owners should establish where they want these necessary functions to take place. Having done so, they must then accompany the dog to ensure that the place he chooses is part of the right area. On the other hand there is nothing like a garden kennel and run where a puppy or dog can be placed during the day without supervision.

KENNELLING

Having kept dogs in many ways during my life I came to the conclusion some twenty-five years ago that the best way to keep dogs was to bring them up to live happily in my own garden kennels or indoors. Our dogs usual routine is go into the kennels for day time or working hours, come into the house during the evening, and then to stay in and sleep well away from a human bedroom at night.

If it is a puppy then teaching him right from the beginning that part of his day is spent in kennels is the best way. For an older dog it may be more difficult, and his acceptance or otherwise of kennels will be affected by various factors. It might be rank, fear or just plain insecurity. The most likely is the latter, so therefore the reader should refer to chapter 6 both in Parts 1 and 2. Certainly money should not be spent on kennelling until it has been established that the dog will accept them or can be taught to accept them. If the dog cannot be left in the garden and is sitting by the back door, barking non stop to be let in, the owner can be sure that exactly the same will happen if placed in kennels.

In such a case the insecurity problem must be dealt with first.

When teaching a puppy to stay in kennels all that is necessary is to put him in the run with toys, chews and definitely water, plus a comfortable bed in a place that is dry and shaded from the sun. A family member he regards as a pack superior and centre of security should stay with him for a short period of time. When that time is reached he should be left without fuss. It is not necessary to say goodbye for this does not happen in the wild.

With an older dog, all of the above things are valid but time should also be spent with the dog close to this run or kennel in a place where he can see, and preferably be able to almost touch the owner. To start with this should be for approximately a five minute session and then be built up by two or three minutes at each subsequent session. This continues until the dog will stop in the kennel for approximately three hours and perhaps go to sleep while he still has company. Once that has occurred, at each subsequent session the person carrying out this association of pleasure with the new kennel should start moving their chair further away on a progressive basis until soon they are out of sight.

If at any time the dog makes a fuss of any description then he should not be spoken to in a soothing manner, nor should any eye contact be made. Furthermore he must not learn that if he makes a fuss he will be let out, therefore unless he has been quiet for 30 seconds or more he cannot come out. With a persistent barker, if necessary remonstrate with him vocally from a distance to obtain a cessation of the noise.

There are many advantages to kennelling. It gives the dog a place of his own where there is no need for him to be continually nagged. If he is chewing it will not be the carpet or furnishings, if he is wet after a walk he can stay there until dry, if he is dirty he can go into his kennel until he has been tidied up again. If guests are to arrive and it is suspected that they may not be completely happy about dogs or their hairs, then he can go in his kennel and run. If it is important for behavioural reasons such as Separation Anxiety that there are periods when the dog is kept away from human company, then kennelling can be ideal.

If the run has a solid floor such as concrete or paving, that is the easiest to maintain. In the UK a grass run soon becomes a winter expanse of mud. Teaching new habits of where the dog should relieve himself becomes easy because that run will be his only options while he is confined to it. It also allows the owner to know what and when he has eliminated from his body.

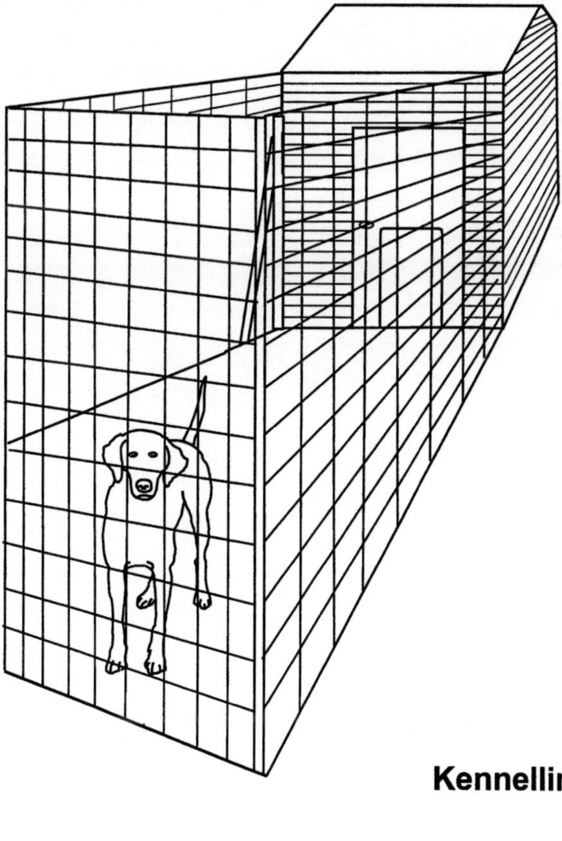

Kennelling

If you have a situation where it is going to very expensive to make your garden secure so that the dog cannot get out, then a kennel and run can be ideal. In the past I have been lucky enough to be asked to go to other countries on dogging business. If my wife accompanies me we have been able to leave our own dogs in their kennels totally stress free as they are quite used to living in them. All we do is ask one of our friendly neighbours to feed and water the dogs, clean their kennels, and take them for walks as well. When we are at home, and occasionally have a heat wave in the summer with the house being too hot for them at night, they can go out in their kennels and keep cool.

Our dogs do not live in the kennels all of the time but while my wife and I are working during the day they are secure from thieves in their padlocked kennels and runs. Then they come in with us at lunch time and stay in for the evening and night.

For a kennel unit it is preferable to have a building that can be walked into and within that the dog has maybe another kennel or a bed raised from the floor. It should be escape proof and of course this also applies to the outside run. As long as the dog can run up and down it can be of any size. The fencing is best made of weld mesh wire and the base of the run concrete. If possible build the kennel and run on a slight incline so that it can easily be hosed down and the water will drain away.

3

BARKING

There are many reasons why a dog barks and therefore the causes should be ascertained before any remedial action is taken.

It is quite wrong to use any device that supposedly guarantees the dog will stop barking without understanding why he barks. For instance, if the cause is a temporary mental illness such as that which can be brought about by insecurity then it would be cruel. Therefore a skilled behaviour practitioner should ascertain the cause and advise the remedy.

INSECURE BARKING

The problem and causes

This is where a dog is left on his own away from pack back up and support and therefore barks, howls or whines in an effort to be reunited. One of the reasons it occurs is because of the dog's level of insecurity.

Solutions

The symptoms are often not just confined to barking but may include household destruction or disruption. It is such a big subject that it has chapter 6 devoted entirely to it and therefore that is where the solution can be found.

EXCITED BARKING

The problem and causes

There are many things that can excite and bring about barking. Perhaps the most common is when a dog believes he is about to be taken for a walk. Sometimes it will be visitors. With others it might be that his dinner is imminent.

Solutions

Psychological: It matters not what excites the dog sufficiently to bark but, inevitably, it will be caused by learned triggers. Therefore certain measures must be taken to confuse any previously learned behaviour that causes barking. First any triggers must be identified and eliminated. That may not be completely possible, so secondly the dog must not learn that barking brings about a good result. If currently that is the case then circumstantial learning of the place, the people and possibly time of day will certainly have brought about circumstantial learning. It does not take long for that set-up to become habitual.

As an example a dog triggered that a walk is imminent may well bark with excitement. If the owner continues with walking preparations and then commences the walk while the dog barks then he will have learned that to bark brings about the desired result. On the other hand, if the owner ceases all preparation for the walk and ignores the dog in every way until he is silent for at least 30 seconds, he will learn that barking does not bring about the desired result but silence does. Therefore silence is rewarded. It's a long process and often very inconvenient but it works. Such a response to any barking that can be identified as triggered by certain actions or words or a combination of both is valid.

Practical: Any attempts at distraction means the dog has the owner's attention, and what he needs to learn is that he gets no response while barking. It is also a fact that to teach most dogs to bark on command it is necessary to also teach them a command to stop. This is because they then understand that they do not bark until told to do so. Because commanded barking is controlled it then becomes easy to stop non-commanded barking by using a learned command. Once again rank is a factor in how easy or not it will be to get the dog to obey the taught command. The middle ranking dog will be easier, while those of high rank may prove more difficult. Advice on how to teach dogs to speak (bark) and cease speaking on command can be found in another book in this series entitled "How to Train A Working Trial Dog". If distraction or teaching the dog to bark and then stop barking on command are not successful, using aversion therapies needs to be considered.

BARKING FOR ATTENTION

The problem and causes

Under this heading I describe a dog who barks at his owner(s) in the house or garden to gain their attention. Generally speaking low ranking dogs do not have the courage to be so bold but they may still find more subtle ways of seeking attention such as nudging with their noses. A dog at least at the higher end of average rank to those much more pushy are the ones who may indulge in this behaviour, especially as youngsters.

Solutions

Psychological: In the first place the dog needs to understand that he is a pack inferior and therefore when he is told to be quiet he does not dispute the owner's right to do so by continuing. If it is a dog that has a long habit of attention barking reinforcement, then maybe some snubbing while he is barking will be appropriate. Certainly no eye contact should be made with the dog, for to do so encourages the dog to continue and teaches that he can make you aware of his barking. Under no circumstances should his quest for attention bring about any success.

Practical: No attempts to distract the dog should be made, for to do so means the dog has gained a positive result. Any practical remedial work should always commence at vocal admonishment level. If that is unsuccessful, and if he will continue to bark while on lead then he can be checked each time he barks. If this does not swiftly bring about cessation then aversion therapies should be tried. Something will work without resorting to technological equipment which I would not advise. When the dog stops barking, then he should be supported so that he knows why the therapy ceases and why he is rewarded

BARKING IN THE GARDEN

The problem and causes

There can be many reasons why a dog does this. Some of them are: Barking at animals or birds he sees or hears. Barking when he hears or sees neighbours in their garden. Barking directed at people and/or dogs passing by. Barking at noises he hears. Fear can be the basis of the problem but so can a strong territorial instinct. If it is animals, birds or people then an over average predatory instinct can cause the problem. Just to complicate matters it is possible that there could be a mixture of two or more of such causes.

One case I had to deal with was two Rough Collies reported by the neighbour to the environmental health officer for barking. This neighbour would not communicate with the owners

Barking at birds

in any way, and if the dogs gave one bark he would open the upstairs window that overlooked the dog's garden and shout at them or hurl things at them. It doesn't take a genius to understand this was provocative behaviour towards the dogs who then barked more vociferously in defence of the territory being threatened.

In effect the neighbour was causing the excessive barking. This not very neighbourly behaviour saved the owner from council prosecution. However, the dogs were successfully taught to stop barking using a combination of noise aversion therapy and psychology.

Solutions

Psychological: It is important that the dog understands he does not have the pack superiority right to bark at anything and that he defers this right to the owners. That alone will not totally stop the problem because on many occasions the owners will not be in the garden when it occurs.

Practical: In the first place the owners should be able to vocally stop the dog barking when he is told to do so. Once again this may mean teaching the dog to bark on command. See page 160 which details where this information can be found. Distracting the dog may have a little effect and will usually only be possible when someone is present.

Almost certainly aversion therapies will be the only likely effective cure for a persistent barker. If his sound sensitivity level is sufficient for noise to be effective then I would suggest that it should be applied remotely. However do bear in mind that the larger the garden the further away from the remotely applied noise the dog could be, and distance diminishes the effects of sound. See more details of remote noise aversion therapy in Part 1, Chapter 8, page 104

BARKING TERRITORIALLY

See Chapter 1 under the heading of "Callers and Passers By".

BARKING AT THE TELEPHONE

The problem and causes

This is a behaviour usually brought about by a trigger of excitement. It particularly occurs where there are one or more teenagers in a family, particularly if they are female.

Any competitive running to answer the telephone will trigger an excited response in a dog easily triggered by this behaviour. It could be a form of hyper-activity, but probably there will be an over average instinct about the thrill of the chase which is, of course, predatory.

Solutions

The advent of youngsters having their own mobile phones will probably reduce the incidence of this behaviour and thus it might never happen in the first place. However, it may not be just telephones that trigger a dog into excited barking, nevertheless the solution will remain the same.

Psychological: All rushing to answer the telephone must cease, with the whole family agreeing to walk to it when it rings. Leadership must have been established, and thus the sole right to carry out the barking should belong to one or more owners and not the dog. When human pack superiority is in place, while he might try it on he will quickly stop when told to do so. However, the word "No" which, in the past, has often been ignored by the dog, will not suffice.

Practical: The psychological advice given above may be sufficient with a lower ranking dog. One of higher rank whose drive to bark excitedly may not so easily respond. If that is the case, remember the story about the German Shepherd Dog under the heading of "Charlie Wyant". This can be found in Part 1, chapter 8 on the subject of aversion therapies. However, as nothing works for every dog another therapy may need to be tried.

BARKING AT GARDEN IMPLEMENTS

The problem and causes

An over average predatory instinct can manifest itself with a dog which chases anything that moves. In all probability that will mean wheeled garden implements come in for the barking treatment, and sometimes the dog will simulate attacking the wheels as well.

Solutions

These will be similar to the previous barking problems, both psychologically and practically. Almost certainly aversion therapies will be required to stop the dog indulging in this unwanted behaviour.

BARKING IN THE CAR

See Chapter 7 - 'Vehicle Travel'.

Barking at mowers

4

DOG WALKING

There are many problems associated with walking the dog. As opposed to behavioural problems confined to a house or garden, many of these can affect other people. It is therefore important that the owners do not allow anything to happen that may be detrimental to others which, at the same time, might harm the image of dogs in society.

NIPPING AT LEGS

This is a behaviour that is often experienced with puppies. Unfortunately, if not curbed it can be a habit which continues into adulthood.

The problem and causes

This is where the puppy or dog bites or nips at the foot, leg or trousers of the person walking him. It is usually a game for the dog and might indicate an over average predatory instinct so that the thrill of going for things that move brings about pleasure.

Solutions

At first it might seem cute or funny in a little puppy but the quicker it is stopped the better.

Psychological: Nothing should be done that enhances the dog's pleasure of his unwanted game or else the principle of 'A = B' is turned on its head. While the owner might still be working at establishing leadership and, dependant upon how much progress has been made, two things can immediately be tried. One is for the dog's target to squeal with pain each time the dog makes contact.

Despite what others write, that reaction to the nip will not work with all dogs. Sometimes a very young puppy will not react in the way we wish to a squeal of pain but will do so months later. Many other psychological factors influence a dog's reaction. Some are the dog's own perception of his rank and that of his target, and then there is the degree of driven motivation to indulge in this behaviour. Another reaction that will work with some is to do the opposite and make sure that the puppy/dog understands that the target is extremely vocally angry. However, if superior rank has not been established this will not work.

Practical: Some dogs might be distracted, particularly by food or toy, but practitioners should be wary of dogs learning that indulging in this behaviour brings about a reward.

I have successfully stopped this behaviour by invoking an aversion therapy of my own, which is to make the action of my feet unpredictable as I walk so that the dog cannot get near to them. It looks quite peculiar to walk like that but it can be very effective. Of course care should be taken that nothing happens which might be injurious to the dog. If really serious, and with dogs more adult than puppy, then aversion therapies can be tried which, when successful, will invoke the 'A = B' principle.

PULLING ON THE LEAD

This heading refers to an adult dog whose behaviour of pulling strongly on the lead is habitual. In some cases the dog will have been to training classes but the habit remains. The problem therefore becomes behavioural rather than a simple training problem. As the advice written herewith is correctional it is not how a puppy or young dog should be trained. Firstly motivational methods of keeping the dog in position should be tried. Then perhaps, with low ranking dogs, stopping and/or turning in the opposite direction will have an effect every time the dog surges forward. While some have success with such methods others do not.

The problem and causes

One of the main factors driving a pulling dog is his perception of the difference in rank between the walker and himself. With a dog having the higher score, the bigger the difference in rank increases the likelihood of pulling and the degree to which it is done. Another factor is the habit that the dog can inadvertently learn of always having to be on a tight lead. Then there is the driven motivational excitement of wanting to get to the lead free part of the walk as quickly as possible, or to the next interesting smell.

However, ranking differences can never be dismissed from this problem, and I have found that the lower ranking dogs are very easy to teach to walk to heel for they have no leadership aspirations to be up front. They are the types of dogs which will defer to their owner by backing past or into them, such as when man and dog approach a door or gate. That is because of the deference a lower ranking dog will often display to an acknowledged superior about the right of who goes through narrow openings first.

Pulling to a post

This is easily proved by observing two untrained adult extremes of rank being walked on lead. Very low ranking dogs will hardly need any training at all while dominant dogs will pull strongly, requiring a lot of work from the trainer.

Psychological: It must be evident by now that much effort should be made to teach the inveterate pulling dog to have a perception of being the lower ranking animal. Therefore everything written in Part 1, chapter 4 applies. However, it is not as simple as that for we have already established that rank can change territorially to non-territorially. If a great deal of psychological work has been put in place in the house and garden a dog can perceive that one or more of the family are his superiors in such confines. Once taking the dog out, all of that can be overturned by the very fact that, practically, the dog pulls on the lead. That means in non territorial situations he has reclaimed superiority.

Once again rank will play a big part in the degree to which that becomes evident. Also one can be sure that, if giving advice on a high ranking or dominant dog, when asking whether the dog pulls when walked the reply is a predictable "yes".

What has usually been taking place is that collar and lead are attached to the dog while still in the house and then he has been allowed to pull his way through doorways and gates. By the time the pavement is reached psychology has become practicality and the dog has asserted his right to be the leader on the walk. Remember the hierarchal social rules of higher ranked dogs having the right to pass through narrow openings first.

Practical: There is advice given by some about every time the dog pulls ahead the handler should stop or turn round and go the other way. The theory being that the dog does not gain success from his attempts to be up front and gives up.

There are some dogs whose behaviour can be changed by this and similar methods but, again, it will depend on the natural rank of the dog and his perception of that of the owners. I have already stated that many motivational techniques attempted will not have worked with this type of dog so other methods must be tried.

Therefore the next things that can be tried are the various dog walking appliances designed to stop the dog pulling. There are head collars or halters and various harnesses. When they work, and they do for some dogs, it does allow the owner some relief from the pulling. However, in my opinion they do not do deal with the cause of the problem which remains simmering underneath, therefore the problem becomes disguised. Also let me state categorically that, despite some claims to the contrary, there is no one device that will work for all dogs. Maybe a combination of a device and the information above will work, maybe it will not. A case of trial and error.

A good practitioner of behaviour modification will always have many possible methods to use. If that person has a solid dog training background then skilful lead handling will almost always bring about a positive result. However, a decision should first be made about the equipment to be used. For me I am quite happy with a plain buckle collar or a half check collar. I have little use for a full check chain but I do not have a problem with their skilful use. When a client's dog is regularly walked on a check chain then that is what I will use for handing that particular dog. What will be important to me is that, with rare exceptions, I require the equipment to be around the dog's neck and adjusted so that, if panicked, he cannot slip the collar and escape across a busy road.

Choice of lead must be decided and usually most leads are far too long for the purpose and need gathering up. Of course a smaller dog requires a longer lead.

There are those who love extending leads but I find them clumsy and not altogether trustworthy. Also there is a little of a fashion accessory about them or keeping up with the Jones' by having something more expensive. Conversely they are also an advertisement to the dog trainer which says of the handler "I cannot train my dog".

As the excitement factor starts the moment the lead is produced in the house, it is a good idea to advise that the lead should be produced 20 times or more a day. This will be for no reason other than to reduce or negate the trigger of an ensuing walk. Next establish a principle that pulling on the lead is unacceptable, and that the handler or owner will go through all doors first rather than be dragged through them by an excitable dog.

If the practitioner is handling the dog then often it is better for the owner to go first, thus inciting the dog to attempt to pull through doorways to get to him. However my experience is that when asking them to do so, unless it is clearly spelt out, they do not watch how the practitioner handles the dog.

I am not a great believer in getting the dog to sit before going through doorways simply because that is not how it happens in the wild. If in the wild an attempt is made to try and push past a very high ranking canine pack member then the likelihood is that his wrath will be incurred. He certainly will not ask the other dog to sit and wait. Therefore I prefer to allow the dog to make the mistake of going ahead of me so that the appropriate riposte can be applied. In fact I want the dog to make this mistake so that I can begin to establish exactly who will be in charge of the walk.

Having adjusted the lead so that there are just a couple of inches of slack in it the two should set off to go through the first interior door. Immediately the lead starts to go towards being tight, and I stress towards as opposed to tight, give the lead a firm check. This is in effect a snatch designed to stop the dog abruptly. The severity of the snatch must be determined by an experienced person who will intuitively know to what degree is should be applied. Such a decision will be taken by knowledge of likely rank plus size and breed of dog.

Inevitably there will be occasions when the dog takes the handler by surprise by lunging ahead, causing the lead to tighten too quickly for a check to be applied before that occurs. To pull it back while taught is quite ineffective and will always remain so. To snatch it becomes impossible. Therefore what must be done is to release the tension by quickly moving the lead holding hand towards the dog then checking it back so that the dog is still checked from a slack lead. Speech or commands are still superfluous at this stage.

Pulling to go out

If previously abortive attempts have been made to stop the dog pulling while consistently using a command such as " Heel", then for this remedial work that command should not be used as it is possible the dog associates it with pulling. This is another reason why any command is best used only when the dog is in the correct position and not when attempting to get him there. With all inappropriate behaviours, if a command is added while he indulges in it, there is the danger that eventually all it does is reinforce what he has previously been doing wrong.

If certain the command "Heel" has not been wrongly interpreted by the dog then it can still be used. Alternatives might be "Close" or "Side". In the odd moments he is walking nicely he can be told he has got it right by the delivery of the chosen command together with praise. In such a way the command becomes associated with the required behaviour, and praise reassures the dog that he is getting it right. Timing of commands is, of course, crucial.

All this will apply as the dog is walked along the street or in the park. Walking should start slowly while ensuring the dog responds appropriately. Think of it like driving a car. The handler must be the driver controlling brake and accelerator. With most dogs this procedure is over and done with very quickly with just a soft occasional check at applicable times being sufficient. Dogs will differ greatly in this respect and the practitioner or handler must decide what is the least amount of check that will still gain maximum effect.

All applicable reassurance to the dog should be vocal without adding touch, for all body language should remain consistent but it is crucial that the dog knows when he has got it right. However, the level of praise may well be critical as praise which is too profuse can incite some dogs back to pulling.

Experiments will determine this. Also some owners want to go over the top with praise. Remember that, as a human, if continually praised every few minutes it will soon mean little to you.

When the dog really starts to get the idea, and this may be after several sessions, a sit at heel should be introduced while walking him. If already taught the command in a domestic situation it will be easy. If not then it should be taught as a separate exercise at home or in the park and not on the walk, therefore avoiding unnecessary confusion.

Whatever happens do not start repeating the command whether or not you think he has learned it. Doing so just teaches disobedience. Give the command "Sit" once, and then wait a fraction of a second before ensuring that the dog complies. The vast majority of dogs learn very quickly that a genuine leader is in charge, and by the time the front door is reached some will already have learned. Again that is dependant upon the rank factor. If when being checked for transgression the dog looks up at the handler's face it is a clear sign that already he has decided that he is not the superior on this walk. What is happening when this occurs is that the dog is asking for instructions. When he does so praise becomes very appropriate. My experience is that some 80% of pullers will do this before reaching the pavement. With the others it just takes a little longer.

Dependant upon how extreme is or was the pulling I will not immediately allow the dog to sniff or toilet. In such a way I am further demonstrating that I, the leader, will determine when the pack sniffs and when they toilet. Having refused consent to do so, subsequently I will give the dog a release command even if he does not yet understand it. Its use is to tell him that now he may sniff and/or relieve himself. He can therefore be allowed to take a few steps towards the appropriate place while letting him have the full length of the lead. Once I have decided that his actions have been sufficient I then let him know he is back under control again by shortening the lead and requiring him to walk close on my left side again. Advice about teaching a release command can be found on page 118 on the subject of returning when called.

There are those with the view that if a behaviour practitioner or trainer handles the dog he is responding to that person and not the owner. Of course that is true, but in the first place it is necessary to ascertain if what is being demonstrated is the right advice for the particular dog under remedial therapy by gauging the dog's responses. Then it is a question of allowing the owner to become the handler while the practitioner becomes the instructor.

Once back home, or before, it will be necessary to explain circumstantial learning, for once the practitioner has been removed from the equation of canine learning the answer could be a little different. Also to be taken into account is that the owner has habitually suffered from familiarity breeding contempt. Suddenly the practitioner becomes like a new school teacher that the pupils, at first, are slightly wary of. All of this must be explained so that owners are aware how skilful they will need to be when walking the dog for the first time without the benefit of expert help and advice.

During this initial session the owner should be allowed to take over as handler so that the practitioner can positively criticise and ensure everything is being done correctly. Also advice about which hand the lead is held in will be affected by the height and size of both handler and dog. If the handler will be more comfortable with the dog on their right hand side then there is no reason why handling on the left side should be insisted upon.

At first do not allow the handler to use a command, for that should only be added when the practitioner's intuition decides the time is right for such vocals to be added.

Walking comfortably

This type of heel training is by necessity firm but not cruel, and is far preferable to a dog which is going to pull and strain on the lead for the rest of his life with the result that the owner constantly nags a confused dog, struggling to breathe and sometimes choking himself. We also have to remember that there are some owners who take the line of least resistance as a means of stopping the pulling by ceasing to take the dog out at all.

Results can easily be obtained in a few days providing that the dog is taken on regular training walks, preferably more than once a day. It will also depend upon the owner's timing and consistency, and if I can impress one thing upon readers then it is this. That old phrase "Give him an inch and he will take a yard" was never truer than with a persistent puller. He should not be given as much as an inch. Half an inch movement forward should be sufficient warning that he is creeping forward from his required position.

If allowed to take much more than an inch before action is taken the handler will be losing the argument. Remember be firm, be fair, be consistent and praise where appropriate and immediately, quickly, or eventually, he will defer to higher pack status.

A spin off from this type of remedial work is that the dog's perception of the handler's rank is generally vastly improved. It is also quite strange that when remedially walking a pulling dog that is also nervous of people, it does not seem to confirm his perception that all strangers are a danger to him. In fact the opposite is the case as so many fearful dogs have shown far less fear to me when I have taught them they cannot pull when I handle them. Of course I have gained their respect but in some peculiar way it goes a long way to gaining their trust. Results can come startlingly quickly removing weeks, months or years of nagging. Now that is what I call kind dog training!

RETURNING WHEN CALLED

The problem and causes

Under this heading I am referring to a dog that while being walked off lead will not return when called or only does so eventually.

Past consequential learning will to some extent influence how the dog responds in an open environment. It can also mean that, while some dogs that are a problem to get to return carry out a good recall exercise at a training club, they may remain difficult to recall when free in the park or woods.

There are various reasons why a dog ignores a call designed to get him to return. It is easy to blandly say that the dog is not trained but that may only be part of the reason.

In Part 1 of this book I have written about how an owner's perceived power over the dog diminishes with distance. I have also mentioned how puppies that return well at home need to be taught to do so in the park before circumstantially learning the wrong habits.

Of course rank plays a part as, when there is a big difference between that of the owners and the higher ranking dog, additional problems are already in place. One other strong influencing factor will be the dog's insecurity level. Those that are highly insecure should be very easy to teach to recall. On the other hand dogs with average insecurity levels may not be quite as easy but should still respond when too far away for their own comfort. Then of course there are the distractions of all other possible occurrences usually encountered when walking a dog that is running free.

Solutions

While the normal indoor dog club method and set up of teaching dog's recalls is valid, for me there is no substitute for also teaching the dog to respond when running free in wide open spaces. This is what I do when asked to solve a recall problem and, in the vast majority of cases, the dog can be taught by making the owner aware of how to invoke his insecurity.

Psychological: It involves the psychology of pack instinct and it is taught without the dog even having a lead on. Dog trainers have long known that walking away from the dog and/or playing hide and seek can be very useful when teaching them to return. I have used that ploy for over 45 years. However, when starting to work with problem dogs I knew I needed to know more about why many dogs would follow their owner and have a maximum distance away from them, which then compelled them to return.

It was applying my mind to the causes of separation anxiety that also gave me the answer to the variations dogs display of willingness to return to their owner when running free. What I found was that all dogs have varying levels of insecurity. For me that word describes the distance that a dog will be prepared to leave the pack (owner) before feeling that he is too far away for their back up.

I realised there would have to be two distances. The shorter of the two being when there is no other factor that is a greater lure than staying within a safe radius of the pack. I call that the dog's 'Basic Secure Radius'. However, when there is something of greater interest that is further away which will invoke the dog's driven motivation to go to it, he will go beyond his basic secure radius to reach it. That means he is temporarily prepared to go beyond the limit of his basic radius. Note that I have stated temporarily, for time away from back-up is another factor. I call this larger area the dog's 'Distracted Secure Radius'.

Dogs are quite good at judging distances, not in metres or yards but in what it means to them. If an owner regularly studies his dog, both in distracted and non-distracted situations, he will find that a figure can be placed on the distances the dog will be prepared to be away from him. Minimum and maximum times away from the pack can also be noted. Those time differences will be affected by the degree of driven motivation involved and the level of his insecurity.

When there is more than one person with the dog they must remain together for they will be the centre of security and there should only be one such centre. Where there are two or more dogs that are regularly walked together it could be expected that one will use another as a security factor. Unless the bonding with the canine pack members is greater than that with their owners, this will not be the case. What is more likely is that two or more family dogs will display differing distances and time away from the owner.

The following diagram shows these radius principles away from the pack's security to better effect than words can.

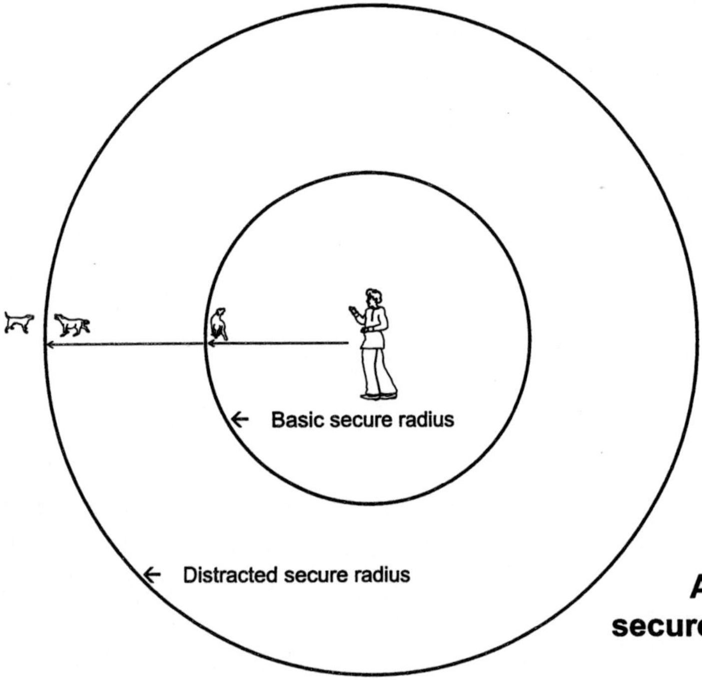

A dog's secure areas

Reading chapter 6 about Separation Anxiety will familiarise the reader with an insecurity scale of one to twelve, with the levels below six describing the least insecure and those above describing higher levels of insecurity. All but a very small minority of dogs with or without a poor or non existent, response to a recall can be taught to respond by this method.

The ease or otherwise in which they will respond to this method will be determined by their insecurity level plus a much smaller contribution of natural rank. Only those almost completely secure and maybe with the combination of high or dominant rank will not respond to this method. Having written that, I have never been involved with one dog that showed no response to this method at all.

Psychological procedure: The dog should be taken to a large area that is safe for the dog to be off lead. Before the lead is removed he should be well away from any possible traffic danger. He must also be taught that the practitioner or handler has a high value food available. Food should be small enough to be concealed under the finger tips of the thumb and next two fingers. In other words it must not be visible or else clever dogs will only respond when they can see the food. Making it invisible means that the greedy dog must respond in case there is food available.

When the dog is at a lead's length of over a yard the hand containing the food should be extended towards him and, as he moves towards it that hand is gradually withdrawn to the groin while backing off continues. When finally he is very close to the hand and groin he should be allowed to try and work the food out of the fingers, and it should be ensured that this takes three to five seconds while the free hand fondles the dog around the neck. This fondling is part of his new learning. It is to become a ritual simulating getting a finger under his collar if necessary, and also teach the dog he cannot grab the food and run off again. Telling the dog to sit is an unnecessary complication at this stage.

Once he has the food he should be given any release command he understands and, if he has not yet been taught one, now is the time for that to commence. I prefer that it is not an over used every day word such as OK. I use "Off you go" but any short phrase will do. Teaching it is quite simply a matter of repetition, for quickly the dog will understand its meaning as it will be connected in his brain with something of pleasure which is being given his freedom again.

If in the past the owner has eventually captured the dog only to replace the lead and take him home he will have learned that returning to his owner was not a good thing to do. Therefore part of his new learning curve will be that returning is doubly rewarding with food and freedom quickly available. At this stage the dog is still on lead and the above procedure should quickly be repeated several times. Once certain that the dog knows food is available he can be given his freedom.

As soon as the dog is free again the pack should walk in the opposite direction to that which the dog takes, but observe him at all times. The object will be to determine when he becomes insecure. This will be displayed by the dog suddenly becoming aware that he is beyond his basic secure area. Usually the dog will have been sniffing and suddenly look up. When he does so and finds his distance away is too great he will make the first couple of paces towards the pack to rejoin them. As soon as this occurs the practitioner or handler should call the dog once excitedly while the pack stay together and, importantly, continue to move away from the dog. This is the first time the dog should be called, and if possible it should not be repeated during his return. Certainly until the dog has been psychologically forced to return, no usually ignored return command should be used as that will only teach disobedience. More than likely this will be what has previously occurred, setting up a habit of owner calls me and I ignore it.

If not already doing so, the pack should turn and face the dog as he closes to the last five to seven yards but still continue to move away by walking backwards. If the dog comes right up to them that is good but they should continue to back off. However, that will be unusual with a dog that has habitually learned how to avoid capture. I call this the last few yards problem for the dog is now secure and has little need of being very close.

The last few yards

Most likely the dog will run past the pack when they should immediately turn round and back off again in the opposite direction without giving additional commands. If necessary that should occur several times but always the hand containing the food will be extended as a lure to get him in close. Eventually the tit-bit conscious dog will find the lure of the food too great, particularly as the pack seems to be doing the opposite of trying to catch him. As he closes for the food the previously described method should be used that taught him food was available. Once the dog has eventually worked the food out of the fingers and the ritual fondling has taken place he is given his release command and the whole procedure commences again.

This psychological method works when the dog is at the limit of his basic secure radius or distracted secure radius. With the latter it is just a question of the dog being much further away. Even though distracted while playing with another dog or anything else, a time factor will come into play. At this point the dog will realise that he has been away from security too long and when he looks up to find that this security is further away than the distance he is comfortable with, he will start to return.

Food for recalls

There will be a temptation to call the dog when he is a great distance away but, as previously stated, it should not be done. It must also be realised that when the dog hears his owner calling it adds back in a security factor that back-up is still going to be quickly available to him. Owners need the courage of their convictions to carry out this method when the dog is far further away than they are comfortable with.

There is a problem when your dog may be annoying another person's dog at a distance from you. You can be seen to be irresponsible by walking away when, in fact, you will get the dog away from the distraction far quicker than by going to fetch him. Doing that only puts the security factor in place again. Finally, never carry out this method in any area where the dog can stray into traffic or get into any kind of trouble. Carry out a quick risk assessment first.

Practical: With the very small percentage of dogs that will not respond to this method the only other possibility is to use a long lightweight trailing line. This needs to be at least twice as long as the average extending lead. With it securely attached to his collar he should be allowed to go away from you and do his own thing, particularly when there is something that interests him. As he reaches the end of the line the handler should smartly check the line so that the dog is brought to an abrupt halt. At the same time he should be called and the line gathered in. If he will respond to food then it should be part of the final sequence as described above. When he reaches you he should be praised and a treat can also be given.

I have to say this not a favourite method of mine. In fact I have never found the need to use it, for with a combination of formal recalls as carried out in training clubs and using the dog's insecurity when running free, my experience is that is sufficient to bring about a positive result.

PREDATORY CHASING

The problem and causes

The title of this problem is self explanatory in that the behaviour refers to a dog that is driven to chase all, many, or some things that move. Generally the faster they move the greater the drive to chase but there can be the usual exceptions to that.

Of course it is well established that many dogs will chase cats, while others content themselves with all kinds of animals when they move. Some will have learned that to run towards a stationary animal will cause it to move, which is what they want, and thus their behaviour is reinforced by the animal's flight of fear. Then there are chasers of bicycles and vehicles and those that chase people, particularly skate-boarders, joggers, or anyone moving quickly.

All of this is caused by the predatory instinct which, when it is over average, drives the problem. The degree of intensity of this instinct will determine the degree to which the dog indulges in this behaviour. There can also be the additional complication of over average rank, and again dependant upon how high that might be may determine the exacerbation level of the problem.

Solutions

I think I first saw a classic statement on this subject written in one of John Rogerson's books. It was to the effect that if an owner cannot stop the dog chasing a thrown ball or toy they would have little or no chance stopping that dog in other predatory chase situations. There will therefore be few dogs with this problem that will not chase something thrown. Whether or not they pick it up and bring it back is another matter entirely.

Psychological: This problem is another classic case of who has the superiority right within the pack to chase anything. Is it the dog or the owner? Therefore leadership issues must be dealt with, plus any residual ineffective habit of the owner's attempts at it.

This problem will inevitably have the added complication of leadership perception changing from that which holds good within the house to that within the garden and even more certainly on a walk. On the other hand it may not change at all if the dog already thinks he is the leader in all situations.

Practical: Almost certainly distraction and vocal attempts to stop the dog will have failed but they are worth trying. If a solid response to a drop command is put in place that could suffice for all but the most driven to chase.

However, we have to accept that skilful teaching of this exercise is hard enough for some competition dog owners let alone pet dog owners. Therefore if all else fails, then checking a chasing dog on the lead or aversion therapies will have to be tried. In all cases being on the lead or a trailing line will be important for the handler will have to be close to the dog to be effective.

FEAR OF STRANGERS OR DOGS

I have written much about this subject in Chapter 1, Part 2, but so far little about how to deal with the problem. For the purpose of advice about rehabilitating dogs I have split the subject into two, for there is fear without aggression which is the subject of this heading and fear with aggression. The latter of the two is dealt with in part three which is devoted almost entirely to the various types of aggression.

The problem and causes

To understand the problems and the causes of fear, once again I refer the reader back to this subject in parts one and three as there is little point in repeating that which has already been written. Suffice it to say that fear is very apparent, and when combined with low rank there is less likelihood of the dog escalating to aggression. That said it cannot be entirely ruled out for even the lowest ranking individual is likely to fight for survival.

Solutions

There is no doubt that the final solution to fear based problems are to socialise the dog to that of which he is frightened. Just telling owners of dogs that they must carry out socialisation is not sufficient help, for fear must be thoroughly understood.

Psychological: Socialisation alone is often not sufficient for a cure. Unless the dog accepts the owner as an undisputed pack superior he will have no reason to obey simple instructions such as walking past that of which he is fearful.

Therefore, like so many problems listed in these pages, leadership must be put in place mainly by using psychology the dog will understand, and Chapter 4 in Part 1 explains.

Fear

Where a dog is fearful of people that he has yet to have been able to trust, it does no harm to reiterate basic principles. Those the dog does not yet trust cannot make a direct approach to the dog without having the effect of prolonging the dog's distrust of strangers.

What is a direct approach? For me it is making eye contact, stretching out the hand in friendship and talking quietly or loudly to the dog. It is what I have previously referred to as the three No's. The worst is eye contact for that tells the mistrustful dog that you have intentions towards him and, by habit, that alone will trigger fear.

If that is followed by the outstretched hand of friendship it will be construed as a threatening arm and hand. Speech obviously directed at the dog will be confirmation of the other two.

We must realise that the whole point of socialising is to teach the dog he has nothing to fear. Such therapy takes a step backwards every time a person inadvertently triggers a fear response in the dog. Far better that they deliberately do not make eye contact, keep their arms and hands close to their body, and do not immediately speak to the dog. In other words it is better if, initially, the dog is totally ignored. By doing this most dogs will sniff them and this allows a dog to check out the stranger who is displaying no body language or sounds that can be interpreted as threatening.

Usually after a few seconds a quiet and calm voice can be used but without eye or touch contact. If that provokes no adverse reaction, probably a light touch can be used. The very last contact should be that of the eyes for it is the most dangerous of all. Initially it should be very brief to gauge reaction. If the dog shows the slightest regression then the time to use the eyes has not yet been reached. On the other hand if that goes well, slightly longer contact can be made next time and in this manner a gentle build up of contact can be made.

If the problem is other dogs that are unknown or not yet trusted then regrettably we cannot control eye contact, but with other dogs on the lead it is possible to control their approach. What the frightened dog will not appreciate is a sudden move towards him and, of course, bold puppies and adults are most likely to make such a move. Certainly we should avoid the dog that shows the slightest aggressive body language or vocal sounds. That will do nothing to prove to the fearful dog that there is no need to be frightened of others of his kind.

In many cases it is also pointless to keep socialising a dog to those he has already accepted as trustworthy. When taking a dog to a club for canine socialisation purposes, circumstantial learning may tell him that after several weeks he can trust all those dogs at that venue. However, almost certainly further socialisation will be necessary in many other circumstances with different dogs that the fearful dog has not yet learned to trust.

It must also be remembered that when the dog is displaying fear there is a danger of reinforcing perception of fear by using soothing words and sounds. This is a difficult area, and my experience is that with some dogs it is valid to do so while with others it is not. Only experienced intuition can decide.

Practical: There is little that needs to be done of a practical nature for a dog that is not showing any signs of aggression with his fear. Careful socialisation to people and animals will usually suffice. However, if the dog pulls on the lead on a walk and generally takes a superior attitude to the owner but is still fearful of the approach of others then practicality comes into play. In other words the dog must first be taught he is not in charge of the walk and therefore does not have the right to take decisions about the approach of others.

Flooding

We could describe the technique of flooding as practical but it is a fine line between that and psychological work. Certainly controlled flooding should be considered and in this case it would mean taking the dog to places where there are so many of that which he is frightened of that it becomes a normal experience rather than an isolated one. In other words he is flooded with people. Shopping precincts are ideal, particularly at a weekend, and also standing outside busy convenience stores with the dog usually means much traffic.

Dog shows and events that allow people to take their pet dogs with them are obviously full of dogs. In these places people are mostly intent on their mission rather than wishing to approach your dog, and this is ideal.

When he is ignored by that of which he is frightened he learns there is no problem. If necessary start from a distance, gradually getting closer to the traffic of people and dogs as the required demeanour of your dog becomes evident.

Generally: Dependant upon the level and type of fear, socialisation may have to be worked at throughout a dog's life. A very fearful youngster successfully socialised by the time he is a year old will easily regress if he lives in an isolated area without further contact with that which had made him fearful in the first place. Some of my clients whose dogs display fear will tell me that they cannot understand why the dog has a problem because they took the puppy to socialisation classes.

Very often just initial socialisation is not sufficient. It can depend on the home where the dog lives. If it is in an isolated position the dog may regress to fear. On the other hand, if the puppy or dog was born with high levels of latent fear, although living in a populous area, some kind of continuing therapy is usually necessary.

5

ELIMINATION

This is the act of eliminating waste from the body. Problems associated with it are discussed here.

EATING FAECES (Coprophagia)

In my early days as a dog owner this problem did not seem to raise its head very often. In latter years it has, and it may be no coincidence that the introduction and continued expansion of complete dry dog foods has a part to play.

The problem and causes

Not all puppies indulge in what will be disgusting to us, but a proportion do. It is usually commenced in puppyhood and if not stopped can be a habit that continues as the pup becomes an adult. No-one is completely sure why this occurs. Some of the theories are modern complete dry dog foods, bitches cleaning up the mess as they would in the nest, curiosity, a lot of boredom, nutritional deficiencies. Often it is confined to their own eliminations; however there are some puppies that will also eat the faeces of other puppies and/or that of other animals.

Solutions

It could be that greedy puppies or dogs will eat anything that looks like food. Having done this once, if it occurs again then it is starting to become a habit. Whatever the cause, the best way to deal with the problem is to remove the faeces before the puppy/dog has a chance of indulging.

Many people just send the puppy/dog into the garden for toilet requirements without even checking to see if he has defecated. I like to know when my dogs have performed this necessary function so that I can be sure they are always relatively comfortable. Quality of stools also needs to checked as one sign of their health.

Common sense: Being with the dog allows me to immediately remove any deposits, and once they have gone there is no chance to reinforce any habit of eating it. Therefore, to break such a habit the owner must always be with the puppy when he is likely to defecate so that his eliminations can immediately be removed. Of course, if this is done while out on a walk then responsibilities as a dog owner should be honoured by removing the mess anyway. I have owned two or three puppies that initially had this problem and it was always quickly cured in this manner.

Practical: Where a puppy or dog has already developed the habit of eating faeces other than his own he is best kept on a lead until the habit is broken. If after a month of being on the lead he still wishes to indulge, then tie him to something while you drastically change the flavour of faeces you will deliberately walk him past. This can be done by using a strong pepper or extremely hot sauce which should be liberally added to the mess. This makes the cure a taste aversion therapy, for what had previously been palatable for the puppy should now be offensive. This therapy should be carried out for several days and he should soon stop what, to us, is a disgusting habit. As soon as it is clear that he has no further edible interest in these deposits he can be let off the lead again.

As with any type of training there is no one method that will work for all dogs. Removing the deposits is most likely to be successful but what might just take a couple of weeks with one dog could take considerably longer with another.

UNCLEAN IN THE HOUSE

Clean on command

One of several excellent aids to training for cleanliness in the house is to teach a puppy or young dog to relieve itself on command. For many years now, to avoid any confusion in the dog's mind, I have preferred to use two different commands for the two motions. So for me urinating is to be associated by the dog with the command "Be quick" and defecation by "Hurry up".

While some success can be had with an older dog asked to learn these commands, the younger the dog is the easier it is for him to learn them. In fact a 7 week puppy will probably do so in two or three days providing the owner is present each time the puppy relieves itself.

It is as simple as watching the puppy or dog carefully and ensuring that he is in an applicable place to perform. Then as he does so keep repeating the chosen command interspersed with good dog etc all the time elimination is taking place. As soon as it ceases the trainer says no more.

It may be that three or four repetitions can be used for a single elimination and this, over a few days, will be sufficient for the puppy/dog to associate the words used. As the puppy or dog learns then repetition of praise is gradually reduced until one command followed by praise is sufficient.

Urination

When and only when certain the puppy knows what it means, the command can be used to ask the dog to perform rather than to wait for it to occur naturally and then add praise to it. Again it will be circumstantial learning and, additional to this, you will be controlling the place or places where you want elimination to occur.

We know from our own experience that defecation on command cannot always take place so the puppy or dog must be watched closely to determine any body language messages that are displayed to indicate that, at present, it is not possible. Currently I have a bitch who gives me that message by stopping sniffing, standing rigidly still and staring at me. This message I have come to completely trust.

The problem and causes

This heading applies to a dog which has either never learned to be clean, or was but no longer is. First it has to be understood that it is highly unlikely that the dog is being naughty but rather that the owner has not yet been able to get the dog to understand the household's requirements about being clean. Therefore the dog continues to act by habit without being aware why the owner shows anger when a mess is found.

Many are such dog owners who believe that the dog knows that he has done wrong for he appears to be sheepish when a mess is found or even before then. If that were the case then it would be very simple to train dogs to be clean. As it is not the case the answers are much more subtle. All that the dog knows from his owners past habit upon finding a mess is that it is a problem. What he is incapable of doing is working back through a series of events about how he came to put it there. We have such brain power, dogs do not.

There are also the dogs that have not performed on the walk and deliberately save it up until they reach home to deposit it in house or garden, much to the owner's annoyance. Once again any retribution against the dog is superfluous for it is another case of the wrong habit having been learned, and that is all it is.

There are some dogs which, having only ever relieved themselves on one surface, will not do so on any other for where they have done it in the past has been perceived by them as a good result. It might be concrete, it might be grass. Then again it might be the carpet in the house. Again that is circumstantial learning whereby the circumstances are attached by the dog to its habit. This can particularly apply to bitches.

Solutions

Psychological: In the latter case above the answer is to take the puppy or dog to a suitable surface that he has so far refused to perform on and stay with him all day if necessary. Eventually nature forces him to gain relief and then the owner should praise and add the chosen command as he finally does so. Generally once a dog has been shown another surface is OK for his bodily functions, a big milestone has been passed. That itself may not constitute a cure to the complete problem, but in similar circumstances to those described it is often a piece of the jigsaw puzzle.

Common sense: Understanding the pattern of a dog's bodily functions goes a long way towards getting a dog clean. When these are known he can be taken to the appropriate relieving place at likely times. Some people leave the food in the dog's dish all day so that he can eat when he wants. This is not conducive to establishing a regular elimination pattern. My own view is that an adult dog should be fed once or twice a day. The food should be placed on the floor and if it has not all gone in 10 minutes it should be picked up and suitably kept until the next feed time.

Owners need to be made aware of the three most likely danger times for elimination to occur.

They are after prolonged sleep. After play or action such as a walk as this shakes everything around and gets the essential parts moving. After feeding is the other danger time. Sometimes it might be immediate but often it will occur within 30 minutes of a meal.

If the dog has not performed during these danger times and he is in the house running free he must be watched like a hawk for those little signs that he is about to relieve himself. Immediately he should be taken out to the appropriate place so that it occurs where the owner wants it to. That is a time to give the dog a lesson in being clean on command by adding the chosen words to his action.

Practical: One of the reasons why man has been able to domesticate the dog is that he is basically a clean animal and will not willingly soil his bed. He may get out of bed and do it not very far away but he does not want to lie in it. Dog people realised some time ago they could use this principle to bring up a puppy to be clean and also to rehabilitate a habitually unclean dog. It is done by confining the puppy or dog to a cage or crate that is only large enough for the whole floor area to be perceived by the dog as his bed. If too large there is the danger that he will decide one end is his bed while the other is his ensuite toilet area.

It should go without saying that the dog needs to be empty before he is shut in his cage, particularly during the night, and if very young maybe the owner should set the alarm so that he can be taken out to relieve himself after no more than three of four hours.

This principle is all about either not allowing the wrong habit to develop or changing an existing inappropriate one. Basically that is all that is usually necessary to get a puppy or adult dog clean. However, it is not just a case of getting a cage and shutting the dog in it while the owners go off and do something else.

Some subtlety has to be employed, and here we revert to psychology, for the animal must be taught that the cage is a safe and secure place to be rather than to immediately learn to be apprehensive of it. This is achieved by initially placing the cage next to one or more of the owners and putting the dog in it with comfortable bedding, and maybe a toy or chew. Ten minutes will suffice for the first lesson, and customisation sessions should therefore be progressively increased.

Once the dog will actually curl up and go to sleep and happily stop there beside his owners for approximately three hours, the owner can be reasonably sure that the dog has accepted his new sleeping quarters.

If at any time he makes a fuss to be let out, then until he has been quiet for at least thirty seconds he cannot be released. In other words he must not be allowed to learn that making a fuss will bring about release from his confinement.

SUBMISSIVE AND EXCITED URINATION

The problem and causes

This problem is when the puppy/dog urinates in the presence of another being. That may be another dog or one or more of his owners. He may roll on his back while it occurs, or be crouching. Often there will be body language signs of submission such as laid back ears, tail tucked under, what some people term as a grin, but it is usually a submissive sign, having nothing to do with pleasure. It is also possible that he will avoid eye contact.

Submissive and excited urination usually occurs in puppies and young dogs and is not confined to either gender. This does not mean to say that older dogs do not do this. I do not altogether subscribe to past theories that it is a sign of submission that the puppy/dog deliberately gives, but rather that it is youth unable to control the bladder when in submissive mode.

Excited urination is similar in that the puppy/dog is excited such as when his owners return but is sufficiently low ranked not to be able to control his bladder. Often the owners will have their legs sprayed as the dog jumps up to greet them. When this type of urination occurs in an older dog then almost certainly the habit factor will have taken hold.

If necessary the reader should refer back to Part 1, Chapter 4 to understand natural canine rank for that is the bottom line of why some dogs are not able to control their bladder when they are displaying submission. In fact a puppy that does this is displaying quite low rank and therefore, amongst other possible indicators of rank, this problem is giving us another type of indication.

Solutions

Psychological: With puppies this problem disappears usually upon maturation or soon after. It is best to grin and bear it and take no action whatsoever for fear of confirming that he should be submissive. Certainly there should not be any scolding or steps taken that effectively chastise the dog for that will be counter-productive.

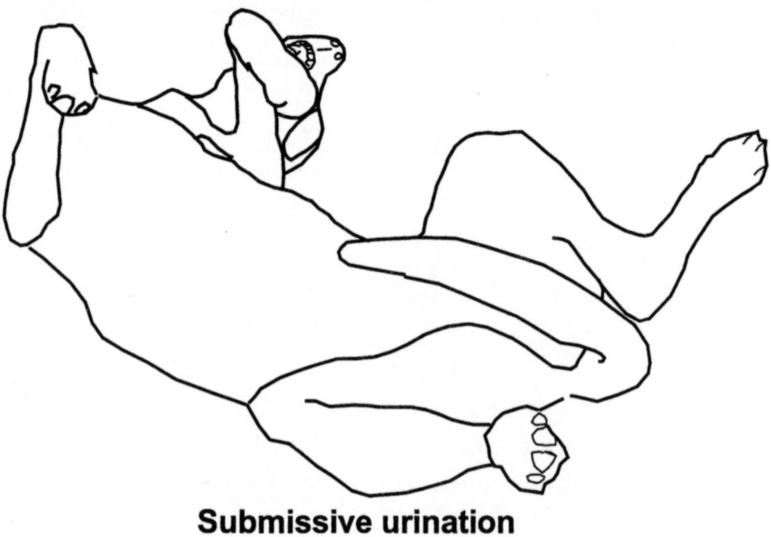

Submissive urination

Practically for submissive: If the family indulge in any game with the puppy/dog that brings about this behaviour it should be stopped. Nothing should be done that reinforces what occurs which might risk making it habitual. This also applies to adult dogs that, habitually, already have the behaviour.

All approaches to those with the submissive problem should be slow and measured and if necessary without direct eye contact. Any hand to fondle them should be directed under the head and definitely not over it. In fact just in the manner described on page 86 of Chapter 1 it is really better to allow the dog to make the approach of his own volition rather than allow people to force themselves on him.

Practically for excited: I can give an example of a dog I owned from a puppy called Whizz. Having come to live with us as a 7 week old puppy, the first morning I greeted him he rolled over on his back and, in excitement, urinated. I immediately realised I had a problem so from there on my morning greeting was low key and casual. That quickly solved the urination problem but what was interesting is the habit it set up thereafter. After 6 months of low key greetings and I decided he must be over the problem, I tried an excited greeting but he remained flat and seemingly unimpressed but did not urinate. Yes, he was pleased to see me, but we had together developed a habit of non excited morning greetings and that at an old age is still the case.

Therefore owners should be taught to work out the right level of greeting for the type of dog they have and not allow other family members to do otherwise. A calm greeting is preferable to great excitement and, for really excitable puppies, maybe upon reunion the greeting should be delayed for five or minutes or more so the excitement factor has substantially abated. Therefore the required habit commences.

MALE PATCH MARKING IN THE HOUSE

The problem and cause

Many are my clients that have had a problem with their dog patch marking in the house who tell me that they suddenly find a wet patch on the carpet or furniture but never catch him in the act of urinating. Usually it is just a squirt such as many a male dog will do on his walks.

Solutions

There are some that can be caught in the act, or the dog can be set up to indulge in what we consider to be a dirty habit but he does not. In either case the owner must try and descend upon the dog, displaying indignant wrath and, if thought to be appropriate, delivering an aversion therapy. Unfortunately the majority will never be seen urinating on the furniture.

Psychological: In the foregoing scenario, in future the dog may use psychology himself as a way of ensuring that the owner cannot catch him in the act again. Therefore the point is there is little we can do psychologically.

Practical: Usually the most practical solution to this problem is castration, as the urge to patch-mark is sexually connected. Some veterinary surgeons will suggest trying drugs first, but if they offer the solution of castration this is the best option to take. However, it cannot be guaranteed as a cure.

6

INSECURITY AND SEPARATION ANXIETY

The problem and causes

Symptoms of this problem are when a dog left on his own without pack back up will howl, whine, bark or become destructive where he is being contained. Additionally extreme stress and panic sometimes causes control to be lost by bowels or bladder. Affected dogs will do one or more of these things.

Since it was first identified much has been written about the problem of Separation Anxiety. I have to say that a lot of the advice given has been ineffectual, seeking to treat the symptoms rather than the cause. Advice such as disguising the signs of impending departure from the home is almost impossible to achieve with an extremely insecure dog. This type of dog will always know the owners are going out and when they have gone, for sure, he knows that he is on his own.

Destruction when left

Other advice that achieves very little is leaving the TV or radio playing to simulate occupied household noises, leaving 'Buster Cubes' and similar for the dog to play or work at. If one accepts that stress is the result of Separation Anxiety then I am unable to understand why a widely suggested cure is to fill a Kong toy, or something similar, with pâté type of meaty food and to give this to the dog when departing the house. The supposed purpose is that the dog will spend all his time trying to extract the food rather than being noisy disruptive or destructive.

As any of these activities are his instinctive way of relieving the stress he is under at the time, unless innately greedy, the last thing he will want is food. If ever we have experienced an extreme panic I am sure that food had no interest for us at the time. So why should we expect food to be the answer with canines. Even if it helps a few, it is still only disguising the cause.

When I first started working with problem dogs I realised that there was a gap in the information available about the cause of this problem. Advice at that time, which still prevails as I write, was that the only thing that caused this condition was excessive company. However, that did not explain why one dog with a certain level of company would have the problem while another dog with the same company level would not.

It did not take long to work out that the missing factor had to be the level of insecurity that each dog is born with. As this varied greatly from dog to dog, as do so many other factors that make up a dog's character, this was the most likely answer.

After coming to conclusions I first published my findings in the Kennel Gazette in 1998 after years of proving them to be correct, as understanding what has caused a dog to become a problem is of great assistance in determining a logical cure.

A dog's dependency upon company can be likened to drug addiction. It particularly applies where the dog has bonded firmly with one or more persons, and the more company the dog has the more he is able to feed upon his addiction and therefore the greater his addiction becomes.

In a typical household, the 24 hour period of night and day can be split into three periods of eight hours. There is the normal working period, the relaxing period and the sleeping period. When someone is usually at home during the normal working period it can be said the dog has eight hours company. Add to that the fact that under most circumstances the dog will have eight hours company in the evening, that means two thirds of 24 hours have been with human company. Add to that again the sleeping period of eight hours by allowing that dog in the bedroom, he then has 24 hours company. All that can be an absolute disaster for the very insecure dog.

Where a family has a member on shift work during the night and sleeps during the day this can possibly greatly increase the dog's access to company if the dog is allowed to be close at that time. With extreme cases the dog may be receiving considerably less than 24 hours company every day, yet still have this problem.

I have attended many cases where the dog sleeps well away from humans at night, yet still suffers anxiety when left. In such cases, when left on his own it will easily be understood how much the dog's anxiety would be increased if he were also to have company at night. However, having written that, in the vast majority of Separation Anxiety cases that I see the dog does sleep in the bedroom.

Not every dog kept under the above circumstances will suffer from Separation Anxiety. Dogs receiving excessive company but displaying no stress when left alone will be hardly insecure at all, accepting isolation without concern. On the other hand, insecure dogs given excessive company are extremely likely to develop this illness therefore it is important that symptoms should be understood. The dog that constantly seeks attention, follows his owner(s) everywhere and quickly returns when called without having been taught to do so, exhibits danger signs. With dogs already afflicted they will display these symptoms anyway.

It is interesting that it is almost impossible for there to be a recall problem with a dog badly afflicted with Separation Anxiety, for the dog's insecurity will always induce a fear of being too far from the pack. How far is too far will vary with the degree of insecurity the dog has. When owners have told me their dog has, in effect, Separation Anxiety and recall problems there is something wrong, for unless the Anxiety when left is extremely low it is almost impossible for the two problems to go together.

Rescue organisations do a good job, but there are those which have a policy of not allowing people to take home one of their dogs because he will be left on his own for too long during the working week. Often this cut off time is in excess of four hours that the dog is going to be without company each day. While such an order may be in the interests of some dogs it is definitely not in the interests of those that are insecure. At the other end of the scale, company is often of little interest to the almost secure dog so to give it to such dogs achieves nothing.

Therefore, the way forward should be for greater attention to be given to the insecurity levels of dogs being re-homed, with advice imparted about how much company is appropriate being dependant upon the dog's level of insecurity. While that would be added work and cost for rescue organisations it is clear that, at present, people taking on what they almost certainly will not know is an insecure dog, can easily be persuaded that much company is a good thing. In effect because of such an impression the dog can be given a mental illness, albeit only displayed when he is left on his own.

Levels of insecurity are very easy to detect with the pet dog. I have already stated that if the dog constantly seeks affection or attention then it is a strong indicator of over average insecurity. If he follows one or more of the family around the house, particularly when there is only one person at home then that is confirmation that he cannot handle life without close back up or support. How quickly and how close he follows is an excellent guide to the degree of insecurity. Such following will seem like total love and adoration but in fact, the dog acting in his own best interests, will be using that person or group of people as the crutch on which he is dependant.

Those big eyes that gaze lovingly at pack members are actually saying, "as an insecure pack member I depend on you and other pack members presence for, without you, I find it very difficult to cope". Therefore, it is not love, as humans understand the meaning of the word but rather dependency.

Such a statement may destroy many people's illusions about how much the dog loves them but it is easily proven. For instance, I have owned a dog that had all the characteristics of being extremely insecure. He often gazed at me in awe and wonderment and, when I moved, he moved with me. However, when I go away for a week or more all that supposed adoration is given to my wife.

Moving from exceptional insecurity towards the other extreme, dog owners will notice that factors such as demanding constant attention, affection, and always being with one or more family members starts to become diminished. With a dog of average security, these traits will not usually be seen as a nuisance, which they can be at the very insecure end of the scale, while the secure dog will not need to be close.

I do not believe there is a dog that is totally secure or one that does not need the psychological presence of any other creature. I may not be correct with this assumption but, if they do exist, they are very rare indeed.

What is often found at the almost totally secure end of the scale is a dog that, whilst he is pleased to see family members when they return home, this display may not continue for the rest of the day as he is content with his own company. Not for him the constant touching and reassurance, in fact the dog may possibly be one which really dislikes being stroked and fussed. He will not be gazing with supposed adoration into family eyes for he has better things to do.. Still he chooses to remain a member of the pack and it is that which tells us he cannot be totally secure, yet he has a low insecurity score.

There are advantages and disadvantages with a dog of over average insecurity. I have already discussed the very big advantage under the heading of returning when called in Chapter 4. Referring back to it and the secure basic radius diagram also gives another strong indicator of how insecure the dog might be. Such an advantage is that it is easy or even unnecessary to teach such a dog to return when called while out for a lead free walk. The disadvantage is, that unless understood and taught to live with his own company, a dog of over average insecurity will of course be prone to Separation Anxiety. How prone the dog is will depend upon his levels of insecurity.

To create a guide of insecurity levels I developed a scale of 1 to 12. Level 1 indicates an almost secure dog that will show little or no voluntary affection and, rather than follow, will be prepared to stray very far from the pack when out on a lead free walk.

Such a dog needs to be taught from an early age to be responsive to a recall command. They will be extremely difficult to teach by invoking such minimum insecurity levels.

At the opposite end of the scale a dog with a score of 12 will be so insecure that in the house or lead free on a walk he will not move more than one or two yards away from his owner despite distractions. Having seen over a thousand cases of this problem I have only seen one dog to which I have given this extreme score. They are the extremes but the reason for the scale being 1 to 12 is that it needs to be divisible by the 24 hours of a day.

INSECURITY LEVELS FROM 1 to12

Levels 1 to 2

These dogs are quite independent. They are more used to providing security for others. If it means going considerable distances from the pack (family) and being out of sight for a long time it is not a problem. They will greet pack members with a wag of the tail but then usually do their own thing, so that in the house, owners may not see much of them. They will be difficult to teach to return when called unless much training is undertaken yet, with skill, some latent insecurity levels might be brought to the surface to assist with recall training. These dogs can be given 24 hours company on an ongoing basis, and when left at home alone there will not be a problem as they are perfectly happy with their own company.

Levels 3 to 4

Dogs given such a score will have some pack security instinct but, while keeping one eye on where other pack members are in the house, it may be some time before they decide to follow them to another room if at all. On a walk they are much more easily distracted and prepared to go quite a long way from the pack before feeling the need to return. They will greet pack members and pack friends after an absence but not overly or for any lengthy periods. They can be taught to return when called by invoking insecurity, but owners will have to work harder at it and realise it may be over long distances. While these dogs are unlikely to be affected with Separation Anxiety, given total company, it is still possible. Otherwise, they will be relatively relaxed when left on their own.

Levels 5 to 6

Dogs with this score will be aware where the owners are in the house at any given time but may not immediately, or always, follow them around the house. They may lie in strategic positions that will enable an eye to be kept on the whereabouts of more than one family member. They like affection and attention but are much less demanding than the following categories. By invoking insecurity they are reasonably easy to teach to return when called, but are temporarily prepared to go further away from the pack when distracted. If regularly given company in excess of 20 hours, separation anxiety can develop, but dependency factors will be far less than the following categories.

Levels 7 to 8

Dogs with these levels of insecurity are very insecure and they will follow their owners around the house but with less persistence than higher categories. They like a lot of attention and fuss. When running free and distracted they will not go further away than the previous level and, when away from security, their instinct will make them aware they may lose the pack if they do not quickly return. They are very easy to train to return when called by invoking insecurity. In fact, training to do so may be unnecessary. Regrettably, given on-going company approximately in excess of 16 out of 24 hours will increase dependency and therefore Separation Anxiety will be induced.

Levels 9 to 10

At these levels we are describing extremely insecure dogs. Constantly seeking affection and attention, they follow owners everywhere, never being more than one stride behind. On walks, they may go a little further away if distracted but are quick to feel insecure and the need to get close to the pack again. When called, in all situations they will quickly return without the necessity of being trained to do so. These are problem dogs as, for this characteristic, even regular human company in excess of 12 or more hours out of 24 can bring about Separation Anxiety when left on their own.

Levels 11 to 12

Completely insecure dogs will score 12. They constantly seek attention and affection and, in the house, follow their owners or a specific household member everywhere, never being more than one stride behind. They will try and go everywhere with their owner, even trying to get in a small bathroom with him or, if excluded, will wait outside to be reunited. On walks, it is unlikely that they will be prepared to stray more than two or three metres, and they are never distracted by anything other than the back up they cannot do without. Therefore teaching these dogs to return when called is unnecessary. If given regular human company in excess of eight or more hours, and sometimes less than eight, dependency will greatly increase therefore, when alone, severe separation anxiety will be the result.

An insecure dog follows

Solution

Domesticated dogs are still pack animals by nature. Even if there is only one dog in a family of humans, he will come to regard the humans as part of the pack that he lives with. Some dogs, particularly those who have thus far had almost unrestricted access to human company for 24 hours a day or even less, can suffer great stress when denied that company. This may be when the family members are all out of the house and it is then, or when they prepare to leave, that stress commences. It usually manifests itself in the form of whining, barking or even destruction of the furnishings by scratching or chewing. Defecation and urination can also be a symptom of this problem when severe.

A common misconception is that the dog is being deliberately naughty when noisy or destructive but, in fact, he is ill and cannot control his reactions. Giving the dog total access to company masks the problem by treating the symptom and only exacerbates the problem when company is impossible. Even worse is that under such circumstances the dog dictates family life by forcing owners to stay at home to provide that company. This illness, while under most circumstances will be caused by separation from human pack members, can also be brought about by separation from animal members of the family. For instance, an older dog that has died. It will largely depend upon whether the dog has had a greater bonding to humans or to other animals, which will usually be canines. However, my experience is that 98% of cases are anxiety when parted from human company even when there are one or more dogs living together. Unless the suffering dog is one of the remaining 2%, another dog for company will not solve the problem.

The dog may look guilty when he has been destructive but that will only be because of the mess. He cannot associate the act of destruction nor his anxiety with the apparent mess and, therefore, any punishment is futile and counter productive. Dogs suffering from Separation Anxiety are not being willful when they become destructive, it is simply that they become agitated and panic when they are separated from human company, with destruction being the result.

To deal with the cause of the problem the insecure dog must be taught to learn to live with his own company. It is therefore obvious that it will be necessary to reverse the amount of company the dog receives and to withhold much of the acknowledgement and touch that he craves. Reducing company is about building enforced separation under controlled conditions.

First it is necessary to ascertain the level of the dog's insecurity to be able to advise the number of hours by which that company should be reduced. Displayed overleaf is a chart which shows the relationship between levels of insecurity and hours of company. With the vast majority of dogs it must be understood that the hours of company that are added to the levels of insecurity will have to be on-going for the problem to develop. The time this process takes will vary from dog to dog but typically it will be beyond six weeks and not more than three months. However, there will be a few dog's that fall either side of this example.

The chart shows that it is impossible to state an exact level of company that causes the problem for all dogs, as it is all about the strength of the two factors which differ. For example, a dog scoring 4 which is only a third of a scale of 12 can still finish up with the problem if given 24 hours company. At the other end of the scale when a dog's insecurity level is the extreme value of 12, just 8 hours company will be sufficient to cause the problem. Then, of course, there are all those in between, and we must not forget possible grey areas.

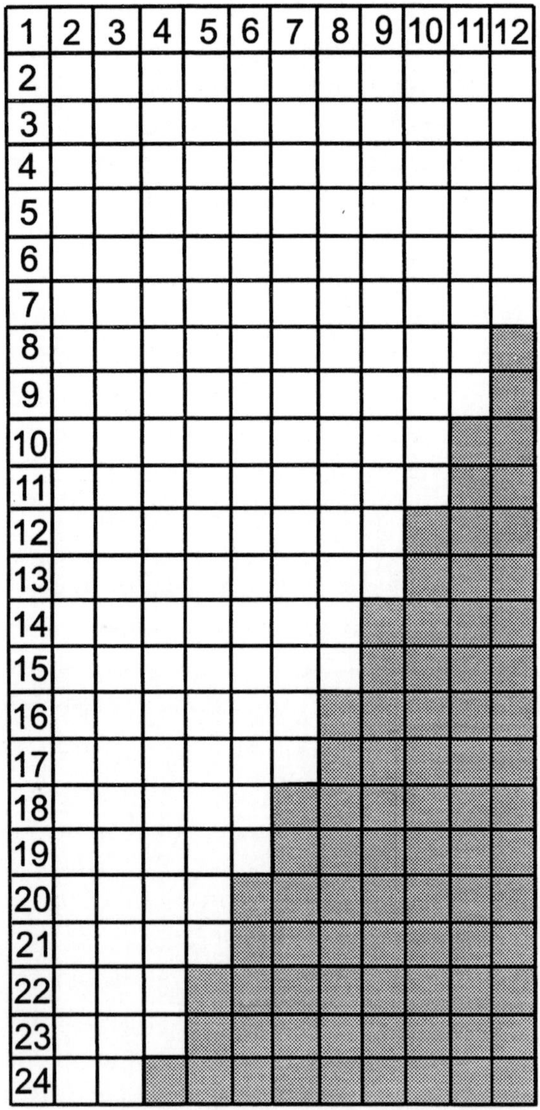

The relationship between Insecurity and Separation Anxiety

Obviously that dog with the extreme score of 12 might be given 20 hours company and we would expect such a dog to display extreme symptoms of the problem. The actual level of each dog's insecurity can be within two figures for a good approximation. With most dogs, as experience on this subject increases, then it is possible to be more precise. However, there will still be some where a good approximation has to suffice.

The first thing to do is to ascertain the amount of company the dog is receiving. This will mean going into the hours each family member is usually at home. As a rule, just dealing with the five working days of the week and ignoring the other two days is sufficient. Having done that, it helps to establish the level of insecurity by using the scale in a backwards fashion. For example if the company level is 16 hours the chart tells us that at the very least the dog's insecurity level is eight, but of course it could be greater. However, using this method gives us a starting point to estimate the minimum hours of company that the dog can be given in the future. If not absolutely sure then it might be the case of advising that company must be reduced by a certain amount, and if that is not soon bringing about signs of improvement then further reductions are necessary. If it can be ascertained that it took the dog a certain number of weeks to develop the problem, then be sure that it will take at least the same amount of time for the dog to be rehabilitated.

Psychological: When able to do so during the day or evening the dog should be moved progressively away from company. This should be to another room or part of the house or maybe to the garden. Starting at two or three minutes duration, over a period of weeks the separation time should be increased in small increments, but only if the previous time lapse has been successful. The dog must not be allowed to think that creating a fuss will force the owner to return. Therefore under no circumstances must a return be made while he is noisy or disruptive. If necessary, he should be told to be quiet from a distance but he must not learn that his actions will bring about reunion. When it is necessary to allow him back, providing that at least 20 to 30 seconds silence has been achieved, he can be allowed to rejoin the owner(s).

If the dog sleeps in the bedroom at night or even on the landing on the other side of the owner's door, that is roughly 8 hours company that needs to be eliminated. When I visit owners with this problem, amongst many other questions that I ask them is where the dog sleeps at night. When the answer is the bedroom, or even on the bed, that is the classic answer and the first part of the problem to be addressed. Moving a dog out of the bedroom without bringing about a stressful reaction is not easy. What will be necessary is to gradually move him downstairs in a two story house or, in a bungalow, a good way from the bedroom. Dog's limited brain power cannot discern that a change is taking place if that change is minimal in a progressive manner. Literally, each nightly movement should be no more than one inch (2·5 centimeters). In such a manner, the dog will be unaware that a change is taking place.

It is impossible to expect the dog to stay where he is told all night or to tie him up so that he is moved by such a small distance. However, this effect can be achieved by using an indoor wire kennel or cage. They can either be made or purchased, and the type used by dog show exhibitors is sold by good pet shops. It must be understood that it will not be cruel to put the dog in such a cage, for it will only be a bed with sides and a top, and if he is introduced to it carefully the dog will come to regard it as his own refuge and enjoy going inside to relax.

Removal by confinement

Ideally the cage should not be much larger than the dog. He should be able to stand, turn round and lie down, but it is a mistake to give him a very large cage and it only makes physical movement of it more difficult. Before using the cage for confinement the owner should ensure that, within his cage, the dog has a comfortable bed which contains his scent. A toy and maybe a chew stick might also help. At first the dog should be familiarised with his new bed for short periods of up to ten minutes. This is best carried out with the cage beside the owner so that the dog associates it with a feeling of security. This security factor will be obtained by reversing the anxiety the dog feels when denied his owner(s) company. The very close presence of the owner(s) will bring about a perception of security. Every now and again, he can be shut in his cage so that time lapses increase and he comes to accept his cage as a secure, comfortable bed. Gradually, over a period of days, confinement periods should be increased, combined with moving the cage further away from the owner. In such a way a gradual increase of time and distance should ensure that the dog happily accepts his new sleeping quarters even though this may mean being unable to be close to his owner.

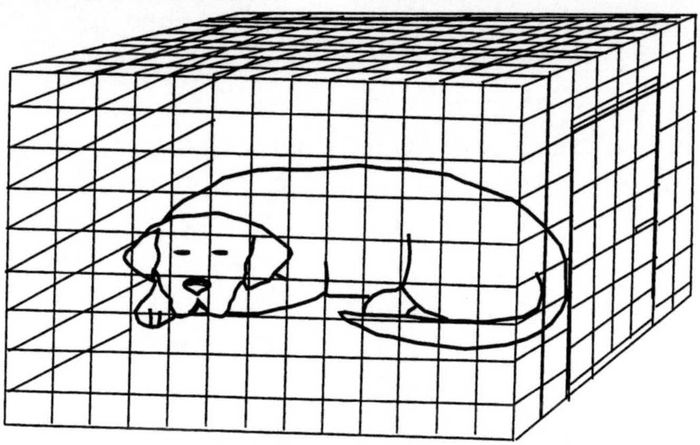

Confinement

As he gets used to this situation the cage, over whatever period of time is necessary, must gradually be moved towards the door of the bedroom. If there are stairs the cage must move in small increments towards them and when, because of safety, the cage cannot be moved the next move will be to the bottom of the stairs. Thereafter it can resume its minimal movements towards its eventual destination.

This process must not be rushed, and should the dog react adversely at any stage, the cage should be replaced in a previously successful position. Eventually this cage should finish up in the newly designated area for the dog to spend his nights. Company, allied to the dog's levels of insecurity, is the problem. As well as reducing the amount of company the anxious dog receives, he should be denied much of the attention and affection that he literally craves for. It has to be remembered that every time he is petted upon demand he is being given a supply of his drug. Most owners think this constant seeking of attention is love and adoration but, in reality, it is the fact that the dog uses his owners as a crutch. This can be proved by asking a friend to dog sit while the owners go out. It will then be found that, under such circumstances, the dog's anxiety level will hardly surface which proves that, in such a case, it is mostly an addiction to humans generally rather than specific humans. Usually this does not mean that the dog cannot have any affection or attention, but rather the owners and not the dog should initiate and control the duration of such interactions. Therefore, when the dog tries to initiate affection or attention it should be denied, but a few minutes after the dog has given up on his quest the owner can sometimes initiate it.

When denying attention, it is better to ignore the dog by every means. If no acknowledgement is given to the dog's presence, he cannot feed his addiction. This means no vocal, physical and, in particular, eye contact with the dog. Where this is not possible then he should be told to go away. With the worst cases, initially all attention demanded must be refused then after a while, providing progress is being made, up to fifty percent of attention demanded time ignored would be a good guide.

Some owners may think that denying attention to their dog takes away much of the enjoyment of owning one, but a caring owner would want to protect their dog from high levels of stress. In any case, while a strict regime may be necessary to start with it need not go on forever as, having cured the problem, the owner will be able to gradually relax the strict regime. This does not mean reverting to keeping the dog in the manner that caused the problem in the first place, but rather find out exactly how far it is necessary to go to control the problem. However, with dogs that are afflicted with this problem it has to be remembered that they will always be addicted to excessive company and, kept inappropriately again, will eventually revert to stress when left.

During rehabilitation, should the dog relapse he should not be admonished. If a relapse occurs then the owners should again examine the amount of company received. It must be remembered that dogs will differ. Those which are only mildly affected will easily be cured, while at the other extreme a severe case may only receive eight or less hour's company to start with yet still be affected with stress when left.

Practical: With the dog removed from the bedroom, human company will have been reduced to a maximum of approximately 16 hours. With more extreme examples such as those with an insecurity score of approximately 9 or above, it will be necessary to reduce daytime company as well. It might be a for a total of four or more hours out of the remaining 16. With someone at home they must teach the dog that he cannot totally depend upon them for security. This is achieved by taking opportunities that fit in with life style to leave the dog in the garden or another room. It may be for 10 minutes here and a quarter of an hour there or by leaving him for the whole period in one session.

When first attempting this, inevitably the dog will react by making it plain he wants to rejoin company. This will be attempted by scratching or noise and, immediately this happens, before he has become stressed, it is possible to teach him to cease by using an aversion therapy. Previously his habit will have been to do whatever he does in such circumstances to attract his owners to his plight. Inevitably, they will have complied with his instructions and allowed him to return to company. In such a way the dog learns how to get a result.

I use aversion therapies in these cases. Either noise or water, and it is uncanny how if one is not suitable the other often is. My preferred method is to use noise as being all enveloping it is easy to use. For this very reason, other animals must be removed from earshot. It is also necessary for the dog not to be able to be totally aware that the owner is outside the door. Therefore those doors with glass panels need to be obscured before therapy commences.

What is interesting, and to a certain extent opposites, when using aversion therapies we are inflicting mild stress on the dog in order to stop him becoming considerably stressed. I find that to be valid as the mild stress will only be temporary and intermittent while major stress of his illness will be over long periods. In other words, it stops them while having the odd grizzle before escalating into a full-blown paddy.

How the dog is taught to desist from his requests to be returned to company is for the person teaching to wait quietly outside the door where the dog will inevitably be close on the other side. At the first sign of agitation, such as a squeak or a gentle knock on the door or greater, that door should be quickly partially opened by the free hand, while the hand containing the can immediately thrusts though the opening to be shaken once or twice in quick succession. As the final shake finishes the arm is withdrawn and the door shut again. All this should take place in no more than two to three seconds and will be the equivalent of a command at least 100 times more powerful than a previously disobeyed be quiet command such as the word "No".

If the dog is relatively bomb proof then noise will not work, and in such cases I use water with a fine jet squirted straight into the face from a household plant spray. The principle remains the same in that it happens quickly so that the dog receives discomfort with the minimum exposure to the company he craves for. A great mistake with these problems and an opposite to using aversion therapies is for the owner to go to the dog to soothe, comfort or even to admonish him. That gives him what he is asking for, which is reunion, even if he is to be chastised.

Noise aversion therapy

As a quicker alternative to removal from the bedroom by small increments, aversion therapies may be suitable. If so, this is carried out by the family getting ready for bed in the normal manner and the dog being placed in the new room he is to spend his nights in. Of course the door should be shut to contain him, and prior to all this taking place a chair should be placed outside the room's door. The aversion therapy apparatus and a good book should be available on or by the chair. At the first sign of agitation, the owner should apply the therapy. No command is necessary as the aversion will do all the talking. It is simply an 'A = B' principle.

When the dog has been quiet for approximately 20 minutes the person carrying out the remedial work can tip toe off to bed leaving everything in place should they need to return. If the dog starts to become agitated again during the night then they should quickly return but only use the aversion apparatus in conjunction with the dog's next call for company. Obviously such therapy must be coincidental with the dog's sounds of agitation.

Some owners seek a compromise by suggesting the dog should be able to stay outside the bedroom door. In my opinion this does not effectively cut him off from back-up, which is what we are trying to achieve. He will know they are just the other side. He will also smell and hear their presence from just outside. Therefore he is gaining some support from being just the other side of the door

Once the dog's habit of insisting upon reunion has been changed then he starts to learn to live with his own company rather than going into self-induced stress. At any time this therapy is used it is all about the dog having controlled times away from company. This together with a reduction in sought after acknowledgement and touch, if applied consistently, will almost certainly rehabilitate the dog.

I have achieved a high success ratio using these principles. I do not subscribe to any of the other oft quoted ways of dealing with the problem for I feel they have little influence on achieving that goal. With the strong belief that teaching the dog to learn to live with his own company is the bottom line, they are only incidental to the problem.

This condition can be cured by patiently following my advice, but it takes time to reverse a procedure which would have built up over a period of time. It could take up to three months, but improvements often become apparent much more quickly. Additionally, while putting the problem right, there is no reason why the very insecurity that was part of the cause should not be used to advantage to teach the dog to come when called. There is no confliction as it is a completely different set up and the circumstances of habit are entirely different.

7

VEHICLE TRAVEL

This is a subject that has become more prevalent as the growth of car travel has increased. While there are a few dogs which have never been asked to travel in these mechanical devices the vast majority do, if only to visit a veterinary surgeon. Some dogs love the car and travel, others are fearful of it. It is this latter category that is dealt with first.

FEAR OF THE CAR

The problem and causes

This heading is to describe a dog which will not get into the car or is reluctant to do so. Usually it is caused by the dog or puppy already having experienced trauma on a journey or as a result of the journey. In the first instance it may be a puppy's first journey from the breeder where his association of fear has been connected to the vehicle. In the second instance it may be caused by a car journey culminating at the veterinary surgery where something happened to the puppy and caused fear. It could be being attacked by another dog in the waiting room or having a needle stuck in him on the examination table. If a dog is attacked in a field it may be reluctant to go into that field again. The problem with fear of the car is similar.

I have dealt with quite a few of these problems in the past. A classic case was a dog that was fearful of getting into the wife's car but would happily get into the husband's. He was away on business most of the time and she could not drive his car. Having asked questions, the answers made the cause easily understood. When the husband was home he took the dog in his car to the woods for long walks. The only time the wife took him out in her car was to go to the veterinary surgeon.

Solutions

Psychological: The dog's perception of fear associated with a particular car must be changed. There are various ways this can be achieved.

Practical: Food is very often the easiest way to bring about a happy association with any place. Therefore until rehabilitation has taken place the dog should have all his meals in the car. Of course this is more easily accomplished with a three or five door vehicle than with a saloon.

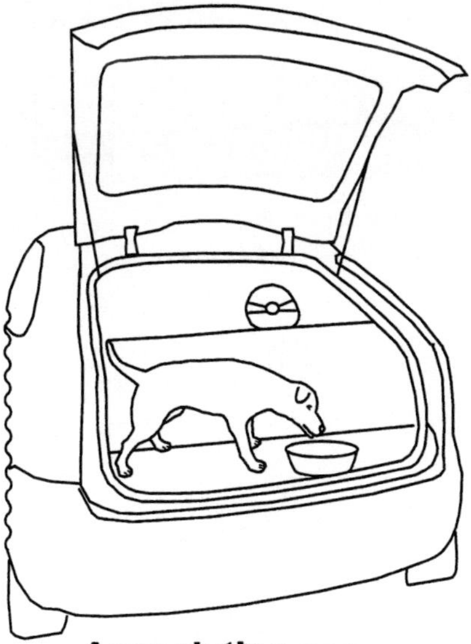

Associating car with pleasure

All other tit-bits given during the day should be discontinued or also fed in the car. If necessary the dog should be enticed into the vehicle with the bowl of food. If he still resists then he must be placed in it. Should he be large enough to avoid being picked up then hunger must become a greater driving force than his fear. In other words the dog must not find he can still be fed in the normal way if he resists.

After a few days most dogs will become more relaxed about getting in, but there will still be more progressive work to be carried out in stages. Owners must be taught to use their intuition about when it is safe to move to a new stage without causing the dog to regress.

Shutting the door the dog entered by commences this process. The next step is for the owner is to get into the driving seat but leave the door open. If all is well, then maybe at the next feed, progress will be to shut the driver's door as well.

Then it will be the routine of seat belts, followed by starting the engine, but not moving the vehicle at all. Next will be the time to start gently revving the engine. When the dog has accepted these additions, then slowly moving away before gently driving around the block would be sufficient for this progression. Obviously everything needs to be slowly built up from that point, but eventually the culmination of the journey must be pleasant such as a walk in the woods or a game in the park. This must occur on many occasions.

What was interesting about the case just quoted was that if the wife got into her husband's driving seat the dog was still happy to get into the car. Therefore the association of fear was with her car. So, with tongue in cheek, it could be suggested that an expensive aid to curing the problem could be to buy a different car! However, unless changes were made to what happens at the culmination of the journey eventually the dog would start to associate fear with the new car. An alterative to this would be to use games with a very playful dog. This is not as easy to accomplish as using food and would require a helper to play with the dog while progression is made, such as driving away.

TRAVEL SICKNESS

The problem and causes

This is similar to the previous problem in that it is fear of the car or its motion that brings about sickness. This might just be salivating heavily, drooling or actually being sick. Often it is confined to young dogs and with careful driving they will get over it. Others need greater attention to detail and because one never knows whether a dog will grow out of it there is the danger of reinforcement of the problem. Therefore all possible steps to remedy the situation should be taken.

Solutions

Psychological: In the first place most of the advice

previously given is relevant to this problem as well, in that a slow and patent build up of the dog's perception of what the car means is necessary.

Practical: The dog should not be fed just prior to going out. During travel fresh air helps, so open the window a little. Sometimes it is the countryside, town or vehicles flashing past that cause the problem. This in itself can bring about nausea. To overcome this possibility it is wise to travel with the dog secured in a comfortable cage, and for it to be covered so that there is no sensation of movement when looking out of the windows. Finally, with stubborn cases there are some drugs that can be effective, and if this needs to be tried a veterinary surgeon's advice should be obtained.

GETTING IN AND OUT

Here I refer to unruly behaviour such as charging in and out of the car. This might more normally be referred to as a training problem but when a behaviour practitioner has been asked to deal with other car related problems, changing the dog's perception of circumstantial learning about who has higher rank in the car might be seen to be part of the cure for unruly behaviour as well.

The problem and causes

Many dogs leap into a car uninvited and worse still jump out in what might be a dangerous situation such as beside a busy road. Usually it will be a case of unruly behaviour brought about by a combination of exuberance and a lack of training.

Solutions

Psychological: On this occasion the psychological aspect will be the dog learning that the owner takes charge of how boarding and alighting from the car takes place. With any other car related problem the dog then starts to understand that circumstantially the owner is in charge of car related matters. On its own, teaching the dog car boarding manners will not cure other problems, but it is all part of the jig-saw.

Practical: Getting In - The dog should be taken on lead to the car and told to "Sit" or "Stand" and then "Wait" so that the handler can easily open the car door without difficulty. If the dog does not yet respond to these commands time should be spent teaching them away from the car. As the door opens the dog must not be allowed to leap in and, if necessary, the command "Wait" should be repeated. If this is not sufficient to keep him still then gentle pressure on the lead in the opposite direction in which he is straining should be sufficient.

Once the door is fully open the dog should not be allowed to jump in straight away, but rather kept still for a further one or two seconds before allowing him to enter, using a chosen command such as "Up" when doing so. If he is unable to jump in because of size or infirmity then he must be assisted into the vehicle. There should be a delay between telling the dog to wait and allowing him to jump in so that he does not subsequently start to anticipate the opening of the door with the command to get in. Once in the car the dog should be secured by whatever is deemed to be suitable. It might be behind a dog guard or in a travelling cage or crate. Alternatively the dog may be secured in a harness or other form of restraint. Having done this the door should be closed and then the handler gets into driving seat.

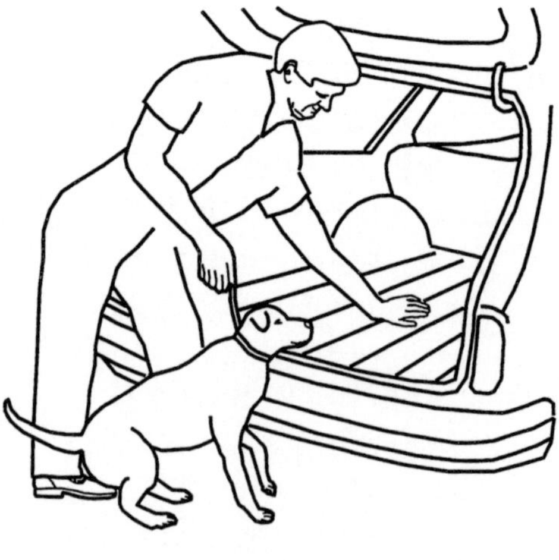

Controlled entry

Getting Out - The door should be opened and the dog freed from anything further restraining him. He must not, under any circumstances, try to lunge out, so the handler should use a "Wait" command again. When the dog is steady the lead should be replaced and if necessary he must be restrained by it. Once all this has taken place, again there should be a brief pause while the dog learns to continue to remain steady before finally being invited out of the car. Once the dog is out he should be told to remain under control in the handler's chosen manner, then the door can be shut.

NOISY AND/OR UNRULY BEHAVIOUR

It is quite amazing how people expect their dog to behave in the car without any training. Of course there are some that do but equally there are many others which need to be taught. The most obvious cure for car problems is never to let them occur in the first place by patiently teaching the dog how to behave.

The problem and causes

Covered under this heading are dogs that just will not settle and pace about or are generally unruly. There are those that howl, whine or bark while the owner is in the car whether it is moving or stationary. Then there are those that lunge at passing vehicles. This latter type will have an over average predatory instinct which drives him to indulge in this hobby. See Part 1.Chapter 3.

Solutions

Psychological: This is another situation whereby the owner might be perceived to have less power in the car than at home. We have already discussed how the perception of rank can change from one situation to another and, with car travel it will usually be because the driver is the owner who, when driving, can do little to control the dog. This lesson is quickly learned by the dog. Therefore the dog must first be taught control exercises in and around the car to establish supremacy of rank in the vehicle.

Practical: For noisy dogs that cannot be controlled by voice alone I have found the best way to alter their behaviour is to use aversion therapies discussed in detail in Part 1, Chapter 8. Almost equally I have found either water or noise to be suitable. Most dogs are transported in the hatch or estate end of a car and therefore I prefer to get into the back seat first. Prior to that the owner takes the dog to the car on lead but waits until I am in it and ready to accept the lead passed through to me. I am then half turned to face what might be a dog guard and ready to apply therapy immediately the dog misbehaves.

In car therapy

If he desists he is rewarded with praise but equally the therapy is reapplied every time he recommences undesired behaviour. It is a simple case of teaching 'A = B'.

NOISY OR DESTRUCTIVE WHEN LEFT IN THE CAR

The problem and causes

This problem is inevitably Separation Anxiety, and in chapter 6 the household version was dealt with. This condition can also become apparent when the dog is left alone in the car. However, that does not have to be the case for some dogs are only stressed when left at home, while others will also have the problem of being without back-up in the car. There are also dogs that will only have the problem in the car and not at home. Which of these possibilities applies will, to a large extent, be dependant upon circumstantial learning, for clearly the circumstances of the car are vastly different from the home the dog lives in. That ties in with my theory that the bottom line answer to the problem is to teach the afflicted dog to live with his own company, but while he may have been taught to do so in one place that does not necessarily apply to another.

When stressed in the car and the dog becomes destructive, regrettably car upholstery is extremely easy to wreck. However, the dog may bark or howl but, whatever way his distress manifests itself, once again he is not being naughty, instead he is temporarily mentally ill. I would suggest the reader refers back to the Separation Anxiety chapter for more detailed information on this subject. The section marked 'Practically' will need to be adapted for use to the car as, inevitably, the use of aversion therapies will be needed. As cars have many windows, then a way of not allowing the dog to know that the owner is waiting outside for any signs of anxiety will have to be found. Otherwise all else will apply, including my point that sometimes it is necessary to temporarily inflict minor stress on a dog to bring about a cessation of going beyond stress when he is left alone.

TERRITORIALLY THREATENING

Before discussing this subject in detail it should be pointed out that most dogs have territorial instincts, and the smaller area they are confined to the greater is there an outlet for that instinct. Therefore this heading only applies to dogs whose behaviour can be described as definitely anti-social.

The problem and causes

Dogs with an over average territorial instinct are likely to bark and lunge threateningly towards those humans or dogs that walk close enough to the car they deem to be their territory.

Solutions

Psychological: Every possible procedure previously described to ascertain leadership should be put in place. This was described in Part 1, chapter 4.

It is important, for the dog has to learn that he does not have superior pack status to assume the right of deterring those encroaching on pack territory. In itself it will not be sufficient and, although possibly having established pack superiority at home, that does not necessarily mean that it is maintained in the car. In fact superiority may have already been lost walking a pulling dog towards the car. If that is the case the pulling needs to be addressed first.

Practical: The reader should understand general territorial instinct discussed more thoroughly in Part 1, chapter 3, and also about aversion therapies in chapter 8 for, inevitably, the only way such a dog will be stopped is through their skilled uses. Dependant upon the situation the behaviour practitioner should either be inside the car with the dog, or outside. If secured in a cage or behind a guard it may be suitable for the back door of the car to be open. As with any corrective remedy the chosen aversion therapy should be delivered immediately the dog indulges in his inappropriate behaviour, and then he should be supported verbally when he gets it right. Timing of the aversion and praise is paramount.

Warning: With this problem and any other it does no harm to write similarly about my warnings in Part 1. Aversion therapies and any practical handling techniques contained within this book should not be attempted with a highly dominant dog or any dog where there is a possibility of being bitten. If the practitioner does not have sufficient skill to evaluate the likelihood of the dog biting they should seek greater expertise.

8

PHOBIAS and OBSESSIVE COMPULSIVE BEHAVIOUR

This chapter title covers compulsive behaviour that afflicts some dogs and is not covered elsewhere in this book. Such behaviour would usually be deemed abnormal, however, before any remedial work is undertaken all possible medical causes should be ruled out by a veterinary surgeon.

These problems can be started by several things. Boredom and anxiety will sometimes bring them about, as will confusion. Whatever the cause of the dog commencing an obsessive compulsive behaviour he will usually gain some satisfaction from it, thereafter it becomes habitual.

AGORAPHOBIA

This is a problem normally associated with humans but I have dealt with quite a few cases of canine agoraphobia. They are not all straightforward and take different forms. Some are easy to cure while others are more difficult.

The problem and causes

So if a dog receives a bad fright which he connects to something that happened outside the safety of his territory, he can easily develop agoraphobia even though walks had previously been the highlight of his day. For instance, fireworks or thunder and lightening that the dog experiences away from the security of his territory can easily bring about this condition. Such dogs will have a greater fear factor as part of their make-up, making it easier for agoraphobia to be added. Therefore the cause is a latent fear plus a bad experience or the perception of such a possibility.

This problem often occurs in the most caring and sympathetic of homes where the owner(s) have been protective and concerned for their dog. It does not have to be noises, but anything that causes fear to be associated with the outside can bring about this condition.

Solutions

Psychological: It is a normal human reaction to try to reassure a dog that there is no problem and coax the dog to go out. When we wish to praise a dog for doing the right thing we quite rightly talk in a soothing voice, and often fuss or stroke him as well. The dog learns that when we speak firmly or in an irate fashion with a lower tone of voice that we are displeased. In such ways we are often able to mould the dog's behaviour.

If a dog shows nervous tendencies and every time the stimulus that triggers the nervous attack occurs we talk to the dog in a soothing voice, and possibly fuss or stroke him at the same time, we are in danger of reinforcing the dog's immediate actions. In such cases it is the dog's fear that we are reinforcing by conveying to him that he is correct to be frightened.

Very often it transpires that the fear of going for walks is connected directly to the people the dog is normally walked by, and that when a stranger takes the same dog out there is an immediate transformation. Therefore before going out a skilled practitioner can gain the dog's confidence and respect as a true pack leader and the dog's reaction to a walk may well be different. When this is the case it confirms that the original occurrence that brought about the fear has either been connected to, or is associated with, the dog's normal walkers. That then teaches us that the psychology of rank is not how it should be within that pack (family).

It is therefore very important that the wrong psychological messages are not transmitted to the dog. Far better that the owner takes the dog out in a matter of fact manner and does not try and soothe, coax or do anything that be misconstrued by the dog as confirmation of a problem.

Practical: One of the first things to try is taking the dog by car to a quiet area where he has never been before. His reaction will determine whether the phobia is associated with the proximity of the house or surrounding streets. If the dog is very playful then, before trying to walk him, a game should be played.

When taken out he should be on the lead so that complete control is maintained. If he sits and refuses to walk, continue walking so that he will have to do so. On no account should he be picked up. If at any time the dog does refuse to walk then, providing he is of a size and weight that makes it possible, continue to walk even though he is dragging behind. This will need to be on soft ground so that there is no damage to the dog's paws. In such cases it is better to say nothing to the dog in case it is regarded as confirmation of his phobia. If and when he starts to walk beside the owner then lavish praise should be given, which should cease immediately should he regress.

Whilst it is not a good thing to make a point of making the dog pass places he particularly associates with fear, neither should they be completely avoided. In other words the owner should not be influenced by what it is felt may frighten him. Avoiding circumstances that worry him may teach that he has every right to believe that the senior pack members are frightened, which will only reinforce his fears. Remember that dogs are experts at deciphering body language and sounds.

It may take some time to effect a change in the dog's behaviour. However, perseverance by a brave and bold example will help, but nothing should be attempted which is in any way injurious to the dog or likely to reinforce his phobia.

FEAR OF PLACES OR AREAS

The problem and causes

Similar to the foregoing is when the dog will go out but has one or more areas that he does not want to enter. Clearly a frightening experience would have occurred in such areas. It is usually noise, but it could also be a place where he was attacked by other dogs.

An interesting example which I remember was a dog that was completely comfortable anywhere in the back garden until a firework party was held next door. On the face of it that sounds like a straightforward case of obvious inducement of fear. However, in this case the dog would remain on any part of the stone patio or path but refuse to go on the grass that he had previously enjoyed and where he relieved himself. Therefore he clearly associated the grass with fear but not the other hard surfaces. That is a classic example of circumstantial learning. A human's comprehension of the problem would be that if he was fearful on the grass he would also be fearful on the patio. It is also an example of the limitations of a dog's mind.

Solutions

Psychological: All of the psychological advice for Agoraphobia applies, and it is important that the walker is seen by the dog as a pack superior. Therefore, if necessary, refer to Part 1, chapter 4.

Practical: Fear inducing areas can also be an area of the daily walk, or it can just be traffic. If it is a continuing frightening occurrence such as heavy traffic or a building site, then the best cure will be what is known as progressive desensitisation. That is when, carefully and gradually, the dog is introduced to the area, starting at a distance. Play, and less likely food, can be used to bring about a new perception of the area. I say food is less likely to be a help as dogs under stress will only eat if their greed drive is extremely high. Therefore if it is to be tried it would need to be high value food. See Part 1, chapter 5, under the heading of food value. Everything should be carried out in a controlled manner as suggested for socialisation. While trying to desensitise the dog, great thought should be given to avoiding anything that might inadvertently reinforce the dog's fear

FEAR OF NOISES

The most common form of this is thunder and/or fireworks. What intrigues me is the hypercritical governmental approach to cruelty where we are supposed to care for animals whether kept domestically or living in the wild. Yet we allow pets and wild life to be exposed to, what must be to them, the horrendous sounds and sights of fireworks.

The problem and causes

It is when a sudden noise or series of them so startle a dog as to make him panic. If he is free he could well run for miles in blind panic to escape the perceived threat to his existence. If he is confined at home, at the least he will shrink away to a corner or shiver with fear.

Fireworks

In some of the worst cases, dogs will try to bury themselves or get under something they think might afford them shelter. Even if never having displayed fear before it will be a latent degree of it from birth that is triggered.

Solutions

Psychological: It is difficult to be pedantic about whether or not any attempts should be made to soothe or reassure the dog. I have already discussed this in Part 1, chapter 6 under the heading of 'Panic' but, basically, an experienced practitioner should decide if it would be appropriate or not with each individual dog.

Practical: When trying to rehabilitate a dog with these problems it is best to use a flooding technique that is combined with progressive desensitisation. To flood in these circumstances is to completely immerse the subject with the fear factor, but if that is done without control more harm than good can occur.

To flood a dog with control it is necessary to obtain a sound effects record, tape, or CD that includes the sounds that worry the particular dog, or even make one yourself. This should then be played daily to the dog for short sessions of no more than one or two minutes at a time. Start at a volume that is barely audible to a dog, let alone humans.

If possible this flooding should be accompanied by something the dog finds pleasant, such as play or being fed. What is being attempted is to change the dog's association of fear with the sounds to that of pleasure. Maybe a long chew at a bone will be something to help him connect pleasure with the sound.

Very fractionally the sound should be increased each day. With minute increases in volume the dog will not be aware of any change taking place and thus continue to indulge in something pleasant while the noise plays. Without doubt the best therapy is play, so it is yet another example of why dogs should be taught to play as puppies.

Working as described should desensitise the dog to sounds he may hear in the house, and it may well help him with the same sounds he might hear while away from the house. If the latter is not the case, then the learning in the house will have been purely circumstantial to those surroundings. In such cases a portable sound playing device with loud speakers needs to be taken to various outside venues, starting again with minimum volume.

Again simultaneously allow the dog to indulge in anything he associates with pleasure. If the sounds that frighten him always come from the same place then that is the area to work towards. If not keep switching locations but never increase the volume other than fractionally for this may be when an attempt at flooding will fail. If a dog is frightened of a certain place because of a bad experience, it is not helpful to drag the dog past that place or to try and encourage him past it. Both actions may only exacerbate his fear, so the only way to tackle the problem is with systematic desensitisation.

Regrettably I have found that unless the dog continues to be accustomed to loud sounds, his fear may return. Therefore, at the first signs of it returning, a fresh programme of progressive desensitisation must be put in place. If a notorious date for such noises is imminent I would advise that the rehabilitation programme is not commenced until that date (or dates) has passed. Several weeks of good work might well be jeopardised by the dog being subjected to massive volume before he is ready for it. Thus regression will occur.

When a dog is frightened he is suffering from extreme stress. All else will go out of his mind other than the fear that becomes all consuming at the time. The most obedient dogs will not respond to their owners under such conditions, and when that occurs it must be remembered that the dog will be in abject panic. Therefore any correction or aversion therapies would be entirely inappropriate.

Dogs can be graded for sound sensitivity on a scale of one to ten. It is unnecessary for me to describe it as it would be self-evident. Suffice to say that the least sound sensitive dog would score one.

FEAR OF DOMESTIC APPLIANCES

The problem and causes

This heading covers all domestic appliances to be found in a house and also those that are used in the garden. Usually the cause is fear which is displayed by the dog's body language and general demeanour.

This can still be the case if the dog is just barking and/or attacking the appliance. There is also the possibility of the cause being an over average predatory instinct, but in such cases there will be other symptoms of this as well such as chasing animals, joggers, bikes and cars etc.

Solutions

Psychological: Readers will not be surprised that again I stress the importance of being seen as the pack leader.

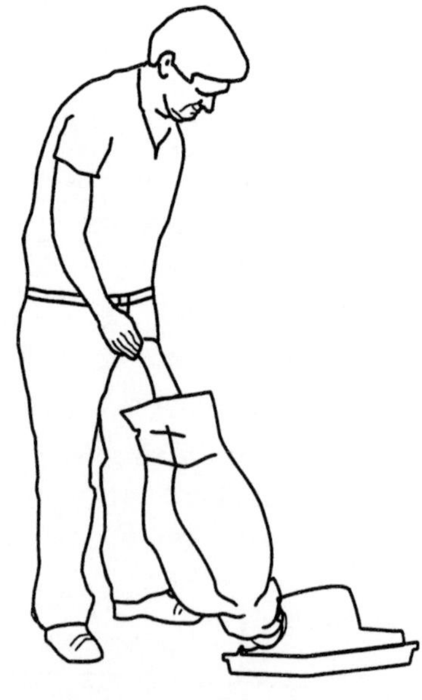

Domestic appliances

Therefore the operator of the appliance should be the only one who has the right to intimidate it. Greater detail is to be found in Part 1, chapter 4.

Practical: Almost certainly at least half of the cause is habitual rather than the original reason. When the dog is being aggressive towards the appliance then it is best to stop this behaviour by use of aversion therapies, and reference to Part 1, chapter 8 gives details. This applies even though the dog may be acting through fear as it is far better to inflict mild stress on the dog than to allow him to go into major stress, which is what happens when he indulges in his behaviour without remedial action being taken.

When the dog displays fear use progressive desensitisation as detailed in the problem which follows. If an owner cannot be consistent with either therapy, then the dog should be removed from the area where the appliance is used so that he is not constantly reinforcing his actions.

OBJECT PHOBIA

The problem and causes

This is when a dog gets a fear of a certain object. Much like domestic appliances, usually these are objects that don't move or make a noise. It might be that something has happened, particularly in the dog's formative months that cause him to be afraid.

Solutions

Psychological: In the vast majority of cases the dog will not act in a bold manner, such as barking at its problem but rather display more obvious fear. In which case the owner should not do anything that might reinforce the fear, such as soothing or reassuring the dog.

Practical: Where fear is displayed then a process of progressive desensitisation should be used by feeding a greedy dog or playing with a playful dog at a distance from that of which it is fearful. Gradually the motivational method can be used progressively closer to the dog but, inevitably, progress must be slow. Should the dog become unsure at any stage then the owner should revert to a previous stage. The idea is to gradually and habitually change the dog's perception of the object from fear to pleasure

CHASING SHADOWS

The problem and causes

This can be an extreme excessive behaviour whereby a dog will immediately concentrate on any moving shadow, quickly trying to catch or stop it. Such a dog will have an over average predatory instinct, but quickly the behaviour becomes habitual. In most of the cases I have dealt with the breed has been Collies whose ability to spot movement and deal with it has been enhanced over very many years.

Solutions

Psychological: There is little psychological work that can be done to counter the problem except to enhance the dog's perception of being a pack inferior. When an owner is perceived to be the higher ranking pack member they will always start with an advantage.

Practical: Only if this habit is very mild will it be possible to distract the dog by playing a different game. The use of food for this purpose will require a dog whose greed factor is greater than his driven motivation to chase shadows. I have yet to come across a case where distraction could successfully be used. I have found the best practical solution is to use an aversion therapy. Almost certainly a dog with this problem will chase the shadows of a torch shining on an interior wall, so it is easy to set up incitement. For further information about the use of aversion therapies turn to chapter 8 of Part 1.

Circumstantial learning must be remembered, for if the therapy is only carried out in one place that may be the only area to which the dog becomes desensitised to his habit. Therefore therapy should be in all areas where the dog indulges in his habit.

OBSESSIVE DIGGING

The problem and causes

This is usually confined to an owners back garden. It has already been dealt with under the heading of "Destroying the garden" in chapter 2. Although not previously described as obsessive, the remedy still applies.

SELF MUTILATION

The problem and causes

When the dog is mutilating its body in some way, such as self-directed licking or chewing, mouthing or excessive scratching, the first port of call must be the veterinary surgeon to check for any medical condition that might be causing the dog to behave in this manner. It may be that the vet advises there is nothing medically wrong or having successfully treated the condition believes that any continuing mutilation is residual habit.

Solutions

If veterinary treatment has been concluded then behaviour modification can be put in place.

Psychological: If the owner is seen as a true leader, a simple "No" will probably be effective. However, with the majority of dog owners, that basic command is unlikely to be anywhere near sufficient to stop the dog, therefore another approach must be used.

Practical: This is a problem where distraction may be effective, particularly if the dog's driven motivation for food or play scores more highly than the desire to work at the part of the body that is being targeted. To gain an initial distraction use anything to make a sudden noise that will gain the dog's attention. For instance roll up a newspaper and knock it down onto a coffee table or such. As soon as the dog's attention has been directed towards the noise, use either food or toy to call him away from where he has been indulging.

The big problem is consistency for, inevitably, there will be times when the dog is left alone. If that is the case and the dog is using his mouth, talk to the vet about providing the dog with a mutilation prevention collar. Much like a lampshade around the dog's neck it prevents the dog gaining access to his body with his mouth.

TAIL CHASING

The problem and causes

Obsessive chasing of the tail can start in a variety of ways. Often puppies will do this as a game, and if the family think that it is cute they will encourage the dog to play the game.

Then there are the dogs which become agitated over something and chase their tails as a form of relief from the agitation they feel. In the worse cases the dog can inflict damage on himself, and this can be serious. Almost certainly, whatever caused the problem in the first place, the behaviour inevitably becomes habitual. Circumstantial learning may play a part, so if the tail chasing occurs in one place or room it is worth checking to see whether it occurs in other places. If such places allow the owner to be more vigilant then the ability to stop the dog immediately he makes any attempt to resume would be valid.

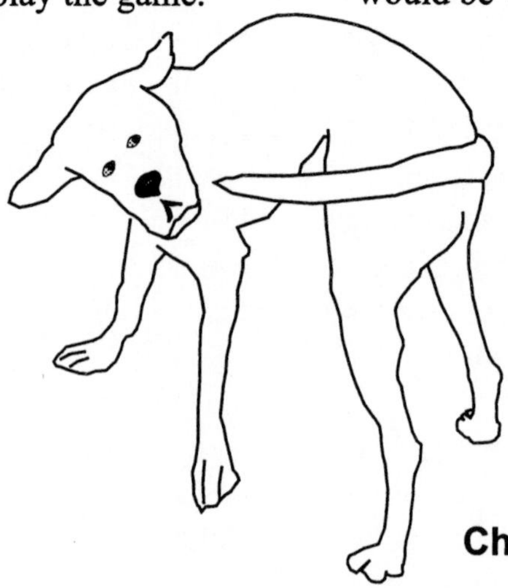

Chasing the tail

Solutions

Psychological: Many owners will try and stop the dog doing this by saying 'No' but the very fact that he continues means that the command is not perceived to have the backup of authority. Therefore some pack leadership work must be put in place but on its own it is unlikely to solve the problem. Every effort must be made to ascertain whether there is a psychological reason as to why the dog is tail chasing. If identified then the cause must first be dealt with.

Practical: As this behaviour will have become obsessive then almost certainly distraction will not work. Therefore I have found it best to use an aversion therapy to teach the principle of 'A = B'. Greater details of this can found in Part 1, chapter 8. Having successfully stopped the dog, he must then be supported and also distracted onto something else.

How to change . . .
DOG BEHAVIOUR

Part 3
AGGRESSION PROBLEMS

1

AGGRESSIVE TALK

Arguably the most talked about inappropriate behaviour in dogs is aggression, yet it might seem strange that we are three quarters of the way through a book about canine behaviour before we reach this subject.

When aggression problems are not split into different types it would probably be correct to say that it is the subject about which most people seek advice. When it is broken down to each different type of aggression then perhaps that is not the case. Suffice it to say that whatever the cause of aggression it is this behaviour more than any other that gives dogs a bad name.

Yet many dog bites are not inflicted using the full strength of the jaw, and the reason when this happens is that the dog is not trying to maim or kill but rather to settle an argument. A dog does not have to bite for me to class it as aggressive, for if it is clear that the current behaviour is aggressive barking that dog might be just a fraction away from escalating to biting the target of his barking. That is a decision that a skilled behaviour practitioner must make and often that decision is intuitive, therefore it cannot be taught.

Unfortunately dogs that do bite do not have the myriad of alternatives that we do in argumentative situations therefore the dog is branded as aggressive when it is often just trying to settle an argument in one of the few ways it has the capability of doing. There have been many cases where dogs have been put down as a result of biting, but sometimes it is purely the fault of the owner or keeper of the dog that has allowed a dangerous situation to develop. People often forget that dogs do not have our sophistication of settling arguments and that their instincts are very basic. When we try to humanise them or compare their behaviour to ours we make a big mistake.

I can count on less than the fingers of one hand the small number of cases that I have felt it necessary to advise there being little alternative other than the dog to lose his life. I am not afraid to do so, but in so many cases of aggression the problem can be turned round. However, it depends on many factors; what the cause of the aggression is, the owner's mentality, the natural born rank of owners and dog, the place where the dog lives and/or is walked, the risk to the owners or to other people.

When giving advice about aggression problems the person doing so must always take account of the risk of the dog biting for the first or subsequent time. Nothing should be done or advised that puts anyone at risk, and surely that should be a basic premise of working with this type of problem.

When someone calls on the telephone having been referred to me by a veterinary surgeon saying "my dog is biting people" then I always immediately ask is that family members or strangers. At that stage all I want is an indication of where the problem lies. If the answer is family then almost certainly it is dominance, whereas if the answer is strangers then the probability is that it is fear. If the answer is neither of those two perhaps the target of the biting is joggers which indicates that it predatory aggression, and if it is the postman then it might be territorial. However that alone will not rule out a mixture of reasons.

Oh, that understanding a dog's behaviour were that simple for, when visiting a client, many questions must be asked before coming to a definite conclusion about the cause, for sometimes initial ideas about where the problem lies might need to be changed. Of course there can be a mixture of reasons as to why the dog behaves as he does.

That mixture might include fear and high ranking behaviour, and in my experience that is not so unusual. Territorial behaviour can be exacerbated by fear and/ or high rank as can predatory aggression. So like all non aggressive behaviour, seldom is the problem black and white for there are always grey areas to consider. Sometimes the cause is straightforward but, whatever it is, the bottom line of why a practitioner in canine behaviour is called in is that the owners want to be able to stop the aggression. I have already stated elsewhere that it is my opinion that roughly half of any inappropriate behaviour from the dog becomes habit driven. Therefore, often but not always, that allows us to take a strong leadership position as part of the cure for the problem's habitual factor. The reason that cannot always be the case will depend on the differences between the natural born rank of handler and dog.

Grooming aggression to owner

For instance, if dealing with an aggressive dominant dog whose born rank far exceeds that of the owners, trying to teach the owners to pull rank to deal with the problem's habitual aspect will probably end in disaster.

In this chapter the reader will come to my golden rule warning about trying to assert superiority over extremely dominant dogs. I have mentioned it in this book before, but to keep pushing this so important message home will hopefully ensure that possibly dangerous mistakes are not made. Nevertheless, other than very high ranking dogs, much progress can be made in stopping aggression when the dog fully understands that his pack position is that of an inferior.

Finally we need to be able to unravel the reasons behind a dog's aggression, and to do so we must have the ability to identify the cause or causes. We should never guess, and that is something I will not do. It is bad enough to guess about the cause when dealing with non-aggressive problems, or also suggest a cure about which one has little confidence or knowledge. Much more irresponsible is the person who does so when a dog's life and the safety of people might be at risk.

2

PLAY FIGHTING

Play fighting is not what we usually regard as aggression. It is a game where these skills are honed in the litter of puppies. I include it under aggression problems mainly as it is a use of the dog's mouth. Also I have seen many clients who believe that their dog is aggressive because it growls in an aggressive manner while they are playing with it or when the dog plays with another dog.

The problem and causes

Certainly play fighting in a litter of puppies is a method of them learning of their ability to defend themselves or an item they have won or currently covet. This may stand them in good stead later in life if they do need defence against aggression. Often we do not need to worry about this type of behaviour, and suggestions given in Part 2 easily deal with what should be a minor problem. On the other hand, if they have been born naturally high ranking or dominant then it may well be something that should be discouraged from the beginning so that when older it does not escalate into a more dangerous activity. Usually it will be very obvious for, at this stage of their life, such puppies are pushy over every aspect of their world.

Solution

When determining whether or not biting is aggression designed to inflict damage, the first thing that must be ascertained is the degree of pressure the dog uses with his teeth. An example is when two dogs are parted from fighting or the dispute has ended naturally. In the vast majority of cases the owners will find there is a lot of saliva covering their dog that came from the other dog's mouth, while there is almost no sign of damage to the skin. This is usually the same with play mouthing an owner's hand.

Psychological: If an owner indulges in rough and tumbles with the dog he needs to understand his dog's body language. Has he used a play bow where he goes quickly down on his front legs but leaves the back side up, usually with his tail wagging. If so that dog has clearly signalled that he has no aggressive intentions. We can also simulate that same signal to our dog, for they understand it and that gives us a chance to signify our intentions. On the other hand, is the dog escalating the game by actions and aggressive noises?

Practical: Generally speaking it is not a good thing to engage in rough and tumbles with a dog of any nature. Such games are more likely to be initiated by males of the family who follow their own instinct in encouraging a macho or male reaction from the dog. With low to underage rank such behaviour will not usually cause a problem. When looking at dogs of high or dominant rank, rough and tumbles are quite likely to encourage aggressive behaviour to develop. If the dog can be truly labelled as dominant then not only does the previous sentence apply to greater degrees but also it can result in developing the dog's instinct to be competitive in finding the best possible position in the pack. Unless certain that the dog is lowly ranked such games should not be encouraged.

In Part 2, chapter 1, I have given advice about dealing with play mouthing and therefore I will not repeat it here except to warn that taking practical measures with a definite dominant dog might well be unwise, and only an experienced practitioner should make such a decision.

3

HIGH RANKING and DOMINANT AGGRESSION

This is a problem often encountered by those working in the field of canine behaviour. It is quite normal therefore any dog aggressive because of high or extreme rank should not be considered to be abnormal.

The problem and causes

If necessary readers should refer back to Part 1, chapter 4 where canine rank is described in some detail for this is the factor that determines those that will or will not be prepared to use aggression to win their arguments.

I deliberately split high ranking and dominant dogs into two categories as for me the latter term defines those that are likely to bite to settle arguments or those that already do so. Therefore dominant dogs score 80 or more on my sliding scale, whereas the category below them which I refer to as high ranking score from 70 to 79.

Often this latter category is described as dominant but I do not see it that way. It is seldom that a high ranking dog will resort to using its teeth. Of course there are always exceptions and my scales are a generalisation. Where we have a high ranking dog which is also exceptionally greedy it may be that using the teeth over issues of food start with a much lower score number, whereas other issues important to it are not sufficiently important to warrant biting.

Then we have the question of how much pressure is exerted by the jaw, and again while there are exceptions to this rule generally speaking the lower the scale number the lesser the pressure applied by the jaw. My experience is that truly dominant dogs are prepared to apply far more pressure to win their argument. In other words they go over the top, whereas if anything lower ranking dogs with a score in the early seventies will apply minimal pressure if they bite at all. Even if not biters they are still very pushy dogs who will easily detect fellow pack members weaknesses and therefore refuse to take instructions over any issue which is important to them. However, the picture becomes distorted when sometimes they will often obey commands but that will only be when they have nothing more important to do at that time. Classic signs of high ranking or dominant behaviour are:-

Non-aggressive signs

(1) Gaining height by getting onto furniture.

(2) Immediately getting on the chair vacated by owner/human he perceives as highest ranked.

(3) Pulling on the lead.

(4) Pushing through doors first.

(5) Complaining or growling when disturbed from rest.

(6) Growling over attempts to remove food or touch dish.

(7) Growling with games of tug.

(8) Growling over attempts to take away guarded items.

(9) Growling when groomed or inspected.

As rank increases so does the likelihood of the above escalate from growling to aggression.

Aggressive action

(1) Over perception that food might be removed.

(2) Being disturbed when resting.

(3) A perceived lower ranked human entering the bedroom he is allowed to sleep in.

(4) Over attempts to take coveted treasures from him.

(5) Body inspected, groomed or towelled; sometimes specific parts of the body.

(6) When a perceived inferior usurps their rights to initiate aggressive action to others.

(7) When others pick up or try and remove toys, even if the dog does not really want them.

(8) When others enter what the dog perceives to be his territory; a certain room, garden etc.

(9) When a perceived inferior tries to stop, or make, the dog do something.

That last item is really a catch all for it covers most issues over which a high ranking or dominant dog will initiate aggression. I have already explained that, when in the company of a dominant dog, I know I will not be bitten if I do not interact it with it in any way. There may be other signs but the above are most common. Also it must be realised that a dog does not have to be classed as dominant because it has one sign, and also a dog with all signs may not be as dominant as another with just a few signs.

Once again nothing is definite and it takes experience to classify a dog as either just high ranking or dominant.

Some people are described as being schizophrenic. I am neither a doctor nor a human psychologist but for me the term schizophrenia almost perfectly describes dominance. In fact a dominant dog has times when he can be described as normal and quite nice, but that will not be so when someone interferes with what he wants, or does not want, to occur. That is when the other side of him will be displayed. Such a dog could be called Jeckel and Hyde, but it would be incorrect.

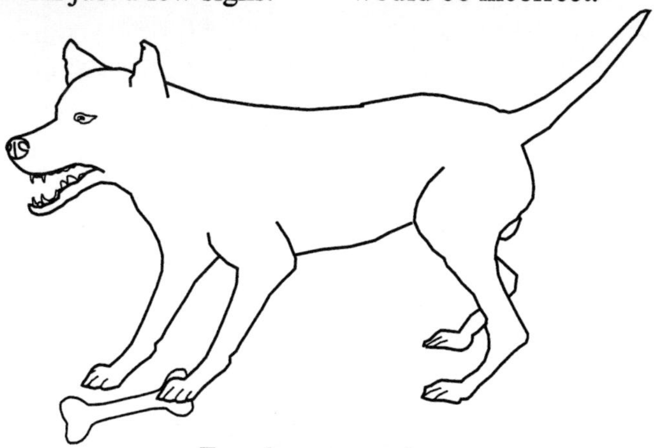

Food aggression

High ranking and dominant behaviour is usually directed at both humans or other dogs. A behaviour practitioner is more likely to be called in when such behaviour is directed towards humans. Where possessiveness over toys or similar is encountered, quite often the high ranking dog will allow humans to play with the toys but not any other dog he lives with. However, an extremely high scoring dominant dog may not allow any being to play with his toys. This also applies to seeing off what are perceived as unwanted intruders. The dominant dog will often accept family members answering the door but may not accept another canine pack member standing level with him barking at the possible intruder on the other side.

When we have dog to dog aggression, if we can rule out fear being the aggressive trigger then almost certainly it will be two high ranking dogs. When they are both male, usually one will decide he cannot possibly win and back down. Almost immediately, or very soon after this has become apparent to the higher ranking dog, he will cease his aggression.

Caller aggression

However, once again this cannot be stated for certain as, for instance if they are confined closely together there may not be sufficient space for peace to be declared.

All of this can be turned on its head if the two protagonists are female and both are highly or dominant ranked. It is quite possible that neither will give way and therefore it becomes far more likely that two such bitches will fight to the death. Again close confinement, such as the back of a car, makes this much more likely to occur.

While it is quite possible for many canine bitches to live together in harmony I have known of several instances where all out war had been declared which would always be continued the moment they set eyes on each other. Who was it once said the female is the deadlier of the species?

So we have the anti-social behaviour of a dog biting others which is always very much frowned upon by society. While I am not one who will defend a dog when it is clearly a public danger, it is a pity that those humans who have the job of pronouncing judgment on a dog do not better understand these issues.

Solution

Other than using psychological measures, determining the solution to high ranking or dominant aggression is not simple. We have two basic practical choices. One is to accept the confrontation that the dog initiates. The other is not to accept it but rather do everything possible to avoid anything that will trigger the dog into dominant aggression. This second option leaves us with a dog that will continue to use aggression to settle his arguments. Therefore that which he is augmentative about must be avoided, while the first option, if successful, will stop or at least limit that aggression from occurring in the future. It is a decision that should only be taken by the experienced.

In my opinion making the right decision cannot be taught either by book, film or lecture as it will be purely instinctive based upon many years hands-on experience. I can merely describe how to try and ascertain which direction to go in, but at the end of the day it can only be a practitioner's skill which can decide.

As a further guide about accepting a confrontation, I can only say that if in doubt don't! If confrontation is accepted and the dog then escalates his aggression which forces that person to back down then they have done far more damage than avoiding the problem in the first place. That damage will have been caused by giving the dog a lesson in how to make the object of his aggression back down. When that occurs more than once it becomes a habitual expectation of the dog.

I deliberately keep repeating my golden rule of a dividing line when a dog will or will not back down, and if they will not do so there is the real possibility of them escalating their aggression beyond that which we can cope with. Therefore if the wrong decision is made people can become injured or something far worse. Therefore I repeat that when making the decision to accept the confrontation it must be ascertained that it will be the dog that will back down and not the human.

That statement does not give someone the licence to be cruel to the aggressive dog, for if that is to be the case they have still got it wrong! If resorting to cruelty to win an argument it is clear that the challenge should not have been accepted in the first place. On the other hand where I talk about accepting dominant challenges then only aversion therapies should be used, for by their nature they are deterrents that have no cruel elements.

Psychological: Quite clearly with an aggressive dominant dog there will be a big ranking problem, and changing the dog's perception of his owners rank can be greatly helped by psychological work. If an owner's natural born rank is only average it would score around the early fifties. If the dog is truly dominant then his minimum score will be in the eighties. Bridging that gap by teaching the owners to act in a much higher ranking manner will be a big task. On the other hand, if the owner ranks in the early seventies or higher then the task becomes easier.

Once again reference should be made to Part 1, chapter 4 so that all the implications of rank are fully understood. In the majority of cases most dominance will be more evident in territorial situations, but can also continue when having moved into non-territorial situations with the dog. To start making changes it is best to do so at home, but we have to remember that may not affect the dog's dominant attitude away from the home.

Every psychological trick should be used to improve and possibly surpass the dog's perception of superiority. Certainly allowing the dog to manipulate even occasionally would not be a good thing. Using snubbing techniques when he attempts manipulation will be very valid, as will be ensuring that the owner is the one to break off any interaction and not the dog.

If games of possession with toys, etc. are played with the dog they must cease unless the owner can retain them until the next game without triggering aggression. Time out or sending the dog to another room when he makes the slightest sign of superiority is also a possible ploy.

Much can be done in a non-practical way, and that will be very important if practical measures are deemed to be too risky. There is an argument for phasing in these techniques gradually, but that will not always be the case and only skilled intuition can make such a decision.

Largely it will depend on just how high the dominant dog ranks. If it is in the late eighties or above then progressively phasing in the dog's perception that leadership is passing to one or more humans may be preferable. If the dog is ranked in the low eighties or below then it may be possible to implement them all at once.

Practical: For me, if I grade the dominant dog higher than 84 then I will not accept his confrontation. Beyond that point is where I am likely to be bitten and I am not stupid. Also I take great pride in only ever having been bitten once.

If the dog ranks below that number, particularly into the seventies which is when I no longer grade the dog as dominant, I will try aversion therapies. It has been my experience that many such high ranking dogs will back down to them. When that occurs then the finest solution of all has been reached because the dog acknowledges another as his superior and ceases to argue when the superior makes a decision.

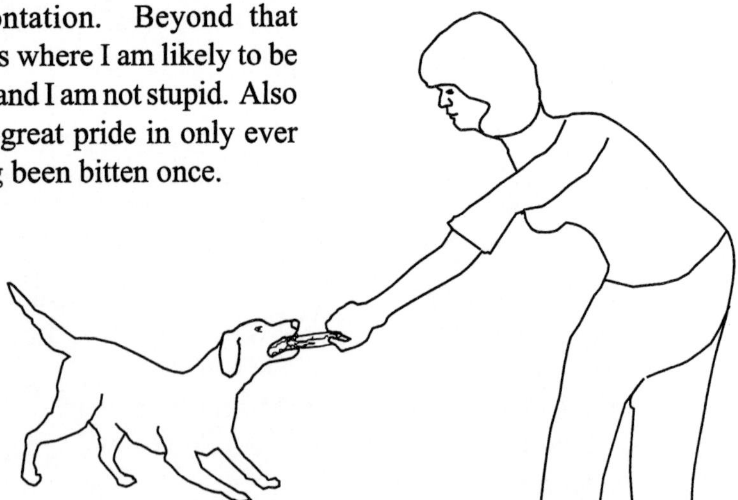

Games of aggression

In this category the dog is not always using his teeth for, while he may have gone beyond aggressive barking, biting may not yet be part of his character. Maybe it includes growling or barging into other dogs and part of a practitioner's skill is to make a judgement as to whether he will escalate to biting and under what circumstances. What has to be taken into account is whether or not the owner(s) will also be able to make the dog back down. Be sure that if a skilled practitioner feels he cannot get such a result then an average pet owner will have no chance whatsoever. Therefore advising on avoidance will be the only route to take. That does not cure but rather only contains the problem.

I have yet to meet a dog acting in a dominant manner that does not pull on the lead when walked. We have to remember that if a dog gets away with such behaviour he is up front as a leader and therefore, non-territorially, the circumstantial learning is probably such that he has even greater pack status than at home. This means that if he is one that indulges in dominating other dogs, and if male with sexual domination thrown in for good measure as well, the owners will have no power to stop the dog's behaviour.

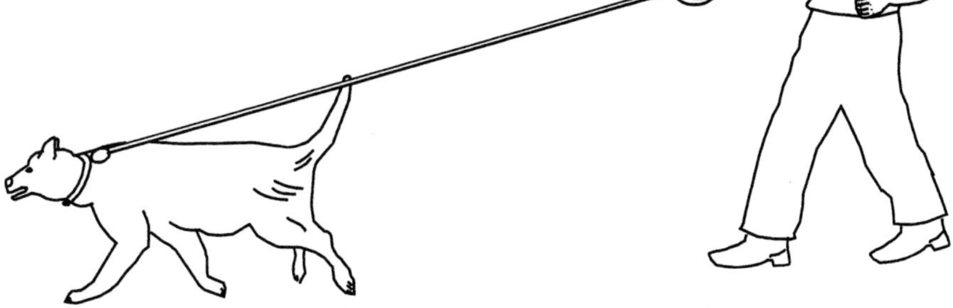

Pulling on lead

Therefore pulling on the lead needs to be dealt with by aversion therapies providing the rule of 84 is not breached. Basically it should be the owners aim to create an aura that only they have the pack status to initiate aggressive action on a walk.

Avoiding confrontations

At home everything that triggers the dog into high ranking or dominant aggression must be ascertained and explained to the owners. When this is carefully carried out it is my experience that they then start to understand the mind of such high ranking dogs. All of those triggers listed under Problem and Cause need to be considered. If the dog is extremely aggressive over food then his food should be prepared without him being present. It should be placed in a room where there will be no humans, such as a garage, shed or kennels, before the dog is encouraged to go to that place. Then he should be allowed access to his food and the door closed. He should remain there until sure that he has completely finished with the dish before being let out again. Subsequently, away from his sight, the dish can be removed.

If the dog becomes aggressive when disturbed from resting then the owners must get him onto his feet in any non-physical way. Possibly that may be tossing food near to him so that he has to get up to claim the food. In the vast majority of cases dogs likely to take action when their peace is disturbed only become aggressive, when lying down. With some it is also when sitting but very few remain aggressive over being disturbed when on their feet. Of course they do not have to be asleep, usually just resting is sufficient for them to become dominant over the issue. If the dog gets onto laps or furniture and subsequently becomes aggressive, then those privileges must cease, for that makes it impossible for the owners not to disturb him. Dominant and high ranking dogs also need to learn that they eat after the family has done so, for the right to eat first is important to the dog whose perception is that he is superior. When food has previously been left available all day that should cease. If necessary the dog should wear a trailing lead, without a handle that could get caught up in furniture. Its use would be to remove the dog from furniture or anywhere he has sought to gain height or privilege.

If he sleeps in a human bedroom, and even worse if that is on the human's bed, then that has to stop. Apart from becoming aggressive when disturbed there is also the problem of the dog seeing, at the minimum, equality of pack status by being allowed to sleep in the best possible place. Therefore the dog must sleep somewhere other than a room where one or more children or adults sleep. Preferably well away from human sleeping rooms.

If he is aggressive over treasures then likely items that he will covet should be removed, and if he does seize anything then if possible totally ignore it. If it must be recovered then the owner should not attempt to take it from his mouth or from between his paws. Subterfuge such as food or toys must be used to distract the dog.
Any other thing that triggers aggression mode will need to be dealt with in a similar manner, for the aim will be two fold. One will be to ensure there is no aggression, the other is stop him reinforcing his behaviour which will already have become habitual.

4

FEAR AGGRESSION

When a dog is reported as becoming aggressive to strangers, that is people, dogs, or other animals that the dog has not learned to trust, then almost certainly the problem will be fear based. That of course does not rule out the possibility of the dog also having other complicating factors such as high rank or any of the subjects listed in this part of the book.

The problem and causes

To understand this subject fully the reader should refer back to Part1, chapter 6, commencing at page 69 and Part 2, chapter 4, commencing at page 186. There the subject of the problem and possible causes were thoroughly discussed. However, to recapitulate, it is my belief that fear is inherited in varying degrees even though it may take an experience learned since birth to make it apparent. Therefore it will take very little to trigger the first fear responses from dogs that have inherited high levels of fear. At the other end of the scale will be those that have inherited very low levels of fear and therefore throughout their life there will be nothing that triggers it. In between these two extremes we have all the other levels.

The problem itself is usually all too easily apparent with the dog backing off, tail tucked under the back legs, ears laid back. However, that is not always the case, for quite often a dog will appear bold by going forward in an aggressive manner when we would normally associate fear with retreating. Of course this may be the result of high rank being part of the dog's character and this is where the cause of aggressive behaviour can become somewhat complicated. Not all dogs that go forward aggressively through fear can be classed as high ranking for sometimes other factors can influence the dog's fear response.

As dogs learn by habit it takes very little for them to understand that the best way of getting the object of their fear out of their space is to go forward. Where this has provided the dog with a positive result, he will repeat that response to fear in the expectation of getting the same result. Likewise the degree to which he does this will not purely be determined by the level of his fear but also the level of his circumstantial learning. If the dog has previously been attacked in a field by another dog, and as a result fear of all or a selected approaching dog that he has not learned to trust has set in. That initial lesson in fear will include the circumstances of that field, or even narrowed to a part of that field, and the degree of the severity of the attack connected to those circumstances. While fear may be greater in that field it does not mean that it will be confined to that specific place.

It takes much experience to be able to correctly diagnose the cause or causes, and fear is a case where the behaviourist can err by making a quick judgement without going fully into the nature of the dog. That requires as many answers to probing questions as the owner can give. Combined with an expert knowledge, much experience and an intuition based upon such experience will usually ensure that the practitioner makes a correct, or at the very least, a close diagnosis of the cause, level of fear and degree of aggression.

In some cases the problem dog will have fear of just humans apart from those already trusted, and then it can be confined to fear of specific types of humans. Often the dog is fine with one gender and not the other, and when this occurs it is usually males that are mistrusted. There are also those dogs that have had their latent fear triggered by the erratic movements of young children, usually because of a lack of exposure to young children during the latently fearful dog's formative weeks as a puppy. Some dogs fear is of people in uniforms or dressed in a certain manner. So the practitioner should never be surprised when a different object of fear is described to them by the owner.

It may just be a specific type of animal, person or object that the dog is fearful of or, in the worst cases, it can more than one. Classic cases are dogs with fear aggression of both dogs and people that he has not yet learned to trust. Some dogs aggressive tendencies are worse when a visitor enters the house they live in, while others greatest fear is passing strangers on a walk. There are many combinations and, while some cases are similar, always be prepared to listen to a different variation on a familiar theme.

Levels of aggression vary greatly. I have already stated that a certain type of barking can clearly be construed as aggressive without the addition of biting and therefore frightening to the person it is directed at. It may be that the dog will contain his fear response to aggressive barking throughout his life. On the other hand he may escalate that into growling and/or using his teeth. One can never be sure but, when asking questions and there has been a clear pattern of his fear response escalating, it would be a reasonable assumption that he will yet go on to higher levels.

There are those that just nip and do little damage providing the nip is not in a vulnerable area such as a face. Some will bite from behind in a typically cowardly way, while others will be much more directly aggressive. Differences in breed will sometimes play a part. A dog with a strong jaw may firmly clamp his target between his teeth. Usually, fear biting is only of sufficient severity to deter that of which the dog is frightened. That is, of course, only a generalisation and many other factors can bring about greater severity of bite.

One might be when the other retaliates, therefore everything escalates, while another may be the interference of sufficient levels of rank.

Solution

There is no doubt that the final solution must be controlled socialisation to that of which the dog is in fear. However, it is impossible to attempt socialisation until the aggression has been eradicated.

We can say that a dog that is intent upon using his teeth as a method of defence is in a highly stressed state. My tried and trusted way of dealing with aggression is to use methods which may place the dog in a mild stressful state. That I find to be valid if it stops stress becoming greater, such as when the dog feels that there is no alternative for survival other than to bite.

Nipping a jogger

At this point I need to qualify method of defence for there are those dogs that started out on the aggressive path because of fear, but have consequently learned to enjoy their aggression. That can be highly misleading for it may well look as though it is more dominant based. Certainly highly ranked fear aggressive dogs will have much of their original driven fear motivation disguised.

Psychological: While I have stated many times the need for the owner to be seen as the clear leader, when dealing with aggression problems that becomes a major must. It is well documented that in a wild pack of dogs the leader will not allow subservient pack members to usurp authority by taking pack decisions on his behalf. We also know that many highly ranked or dominant dogs will not allow perceived inferior canines to bark at possible intruders, reserving that right unto themselves. Therefore the aura of an owner in charge of the aggressive dog both territorially and non-territorially must be that of the undisputed leader. One who has the sole right to bark at anything, the one who will instigate any aggression. In many cases of aggression, when that aura can be established the aggressive dog will either not be aggressive at all or will first look at his owner's face for permission to initiate aggression. Such body language can tell us much.

What must be understood is that there is a level of fear aggression where establishing leadership alone will work or just mild application of the principle of 'A = B' will suffice. Once again we could have a sliding scale of fear, for as the number on the scale increases so the dog's response to the owner or what the owner is attempting decreases.

That does not necessarily mean that there comes a point at which the dog cannot be stopped from aggression but, if necessary, that possibility must always be faced. What is more likely to determine the dog's number on that scale is whether he is mildly, averagely or highly stressed or has gone beyond all that into blind panic. Certainly at the high end of the scale, and where panic is obvious, all normal reasoning of the dog's brain ceases. That means he will take no notice of his leader or any attempts to stop his aggression. This also applies to humans.

Nevertheless, before any dog is allowed to get to the state of panic, total leadership should have been established to give every chance of the dog being curbed from escalating his emotions into panic. Therefore I refer the reader once again to Part 1, chapter 4.

Practical: For my part, when handling fear aggressive dogs I have them on a lead. Once again this is where skilled handling experience with dogs plays its part. I will want to see attempts to go into aggressive mode so that I can determine how to change the dog's behaviour.

There are those teaching that behaviourists should only counsel and not handle problem dogs. Inevitably such people have limited or no handling experience at all and therefore they deem it pertinent to give such advice if they are unable to handle dogs themselves. With such a ploy their shortcomings are not advertised.

When working with aggressive dogs the practitioner needs every possible tool, for in some cases it can be a question of saving a dog's life from compulsory destruction when often that dog's behaviour can be eradicated. Therefore skilled dog handling experience which breeds confident handling abilities should be a prime part of dealing with fear aggression.

What then are the ways in which aggression can be stopped? For me there are three possible practical methods that I will use. One will be checking the dog on the lead. The other two are aversion therapies of noise or water.

Checking on the lead

This technique is no different to that written about in Part 1, chapter 4, where I described how to teach the inveterate lead pulling dog from doing so. It is exactly the same. For instance, in the vast majority of cases the dog will pull or strain forward with aggressive barking, lunging or all the varied ploys he can use to be aggressive to the object of his intentions. That will usually be humans or animals, and therefore the manner in which the lead is checked is identical.

If the dog is also a puller on the lead, before meeting any possible target that problem should be dealt with for in itself the practical application of stopping the dog pulling establishes superiority. When that has been successfully dealt with, some dogs that normally display aggression will not do so on that particular walk. The reason for this is clear in that they have decided they are being walked by a pack superior, albeit a temporary guest of the pack. When that occurs it appears to be quite miraculous but immediately backs up everything the owner should have been told first about what may well occur when working practically. This has happened for me on numerous occasions and when it does it totally proves my theories about establishing leadership. While proving to the owner that the dog can be stopped from aggression, exactly how the practitioner has achieved that must be fully explained and taught to the owner. Again it is about experienced handling.

However, not all dogs will immediately decline to initiate aggression once leadership has been established. It will depend upon the dog's driven motivation to be aggressive and the applied preventative motivation of the practitioner or handler. When the dog first spots that which triggers his fear he will display body language and/or vocals that his normal aggression is very quickly about to commence.

The handler/practitioner should immediately check the dog back or use another aversion therapy as a warning. If that works then quiet brief praise should be used to reassure the dog. If he then indicates that he will have another attempt to become aggressive the treatment should be applied again.

If I generalise I could say that 33% of fear aggressive dogs will not even start to become aggressive when they understand they are being walked by a superior, and another 33% will back down when further demonstrations of superiority are effective. That leaves 34% of dogs. Allowing for 1% as incurable, which from experience is a high figure, we are left with another 33% for whom these techniques will not work. That means we need a different or additional approach to the problem.

Noise aversion therapy

This will only work with dogs that are sufficiently sound sensitive. In the final chapter of Part 1 of this book under the heading of aversion therapies I described the use of a rattle can. I stress some important points such as sound being diminished by distance and the amount of volume to use. I always commence with maximum volume unless I know that the subject of this therapy is extremely sound sensitive. The reason is to ensure that the dog understands that the can is a big problem and one greater than his desire to use aggression to remove the object of his fear. In effect administering the noise creates mild stress but stops the dog going into high stress or panicked biting.

It must be remembered that we can control the amount of volume by the severity of the shake, and we can also diminish the sound by using a rattle can with a smaller volume of space and smaller stones such as path gravel. There is an argument for using the can right behind the dog's tail so that he turns away from his target to check what created the noise behind him.

When this is so then he should immediately be praised for turning away from his target, but timing is crucial, particularly if he is to quickly try and confront his target again.

The alternative is to shake it in front of him to stop any forward progress, and again he is praised immediately he backs down from further aggression. Usually that will be instantaneous. If necessary the treatment should be repeated two or three times at the first target but, and it is a big but, if it is necessary to shake the can close to him more than four times then the practitioner must consider whether this technique will work with that particular dog. It is really as simple as that. There is little point in continuing to shake a can when the dog is not backing down to it for the 'A=B' principle (Part 1, chapter 8) is clearly not effective. When that is the case then water aversion therapy should be tried, but this is more difficult to use because water needs to be directed at the dog's face.

It might be that either noise or water is used in conjunction with lead work, and this is another instance where sound handling experience comes to the fore. Of course if there is any other aversion therapy that is suitable for that dog its use should be considered. However, I have always found that either checking on the lead, noise or water will be the answer to stopping aggression.

It must be understood that dogs will react differently when subsequently walked by the owners alone. At one extreme the owner will have to apply the therapy for some time. However, the other extreme is that the dog immediately backs down when mild use of the therapy is applied. Also remember that where noise is concerned very sound sensitive dogs will quickly react to being shown a rattle can that remains quiet. When that happens the can has become a sight deterrent.

What is very important is to impress that the work is not

finished just because the dog has stopped his aggression. Fear will remain, but such a dog can now gradually be socialised under controlled conditions. It must be very gradual, and to start with the dog may not initiate aggression with his previous target type when a certain distance away. If trying to rush the procedure by exposing the dog closely to the object of his fear too soon he may regress. Of course as this happens the therapy should be re-applied but the owner should learn that he has progressed too quickly. In this case socialisation is all about proving to the dog that he no longer need fear whatever has previously triggered his aggression.

5

PREDATORY AGGRESSION

All dogs have varying levels of the predatory instinct which determine how strong is the dog's desire to chase anything moving quickly, or something that he has already learned will move quickly if chased.

The Problem

I have written about predatory chasing in Part 2, chapter 4, page 185, and suggest the reader refers to that first for it deals with non-aggressive chasing. With some dogs such chasing remains non-aggressive during their life or until stopped from indulging in the thrill of it. With others there can be the addition of biting or trying to hold that which it has caught. That has to be considered to be aggressive and therefore this part of the book deals with the predatory aggression side of chasing. It may be just one type of creature or a variety that the dog chases and, while the obvious will be other animals, sometimes it is just birds. While predatory behaviour towards animals is bad enough, society tends to become more concerned when it is directed at humans. Into this category come joggers, runners, cyclists, push chairs, skate boards, motor vehicles, mechanical devices or anything that moves over the ground.

Dog chasing

Solution

A complication of this problem is when a dog is also highly ranked and, as usual, the higher the rank the more difficult it becomes to change this behaviour. Certainly any games of chase must cease, and that includes throwing items for the dog whether or not he brings them back. Such games only reinforce the dog's predatory instinct.

Certainly, where possible, both psychological and practical methods should be used to establish pack superiority on a walk or indeed in the garden if the problem also exists there. Without the dog having a perception of the handler being the undisputed leader then nothing will work.

When discussing predatory chasing in Part 2, I referred to the ability of the handler to stop the dog in mid flight if something has been thrown for him. If that is possible then that game should be continued to reinforce the stop.

Psychological: Establishing in the dog's perception that the handler has non-territorial superiority will be vital. I must therefore reiterate that if superiority is apparent in the home and garden, that state of affairs does not necessarily transfer to walking the dog either on or off the lead. That will be because of previous circumstantial learning or the lack of it.

Practical: If the dog's driven motivation to chase outweighs any desire for food or play, then attempting to use either types of motivation as a distraction will not work. However, it should be attempted if there is a chance of it working. On the assumption that it is not the case that leaves aversion therapy. In this case, as the nature of chasing means the dog moves quickly forward, water aversion therapy cannot beproperly applied into the dog's face. However, there is always the possibility of using an assistant standing ahead of the dog.

My preferred therapy for this problem is noise. It will often work when delivered very close to the errant dog but, as previously discussed, the practitioner will come across dogs that do not respond in the desired manner. In such cases then checking the dog on a long trailing line would be an option to invoke the principle of 'A = B'. So the chase would be 'A' and the check 'B'. Dogs that are extreme predatory chasers using their teeth when they catch their prey can finish up losing their lives if not stopped, and sometimes a short programme of such therapy brings about the desired results.

6

TERRITORIAL AGGRESSION

The milder form of this problem has been discussed in Part 2, page 127. Of course all of that still applies.

The Problem

This is not uncommon as to a certain extent the vast majority of dogs exhibit some form of this behaviour, albeit in mild terms. It is when there is physical aggression with it that people really become alarmed.

There are so many different situations where dogs indulge in what inevitably becomes a habit. For instance the dog may roam free on a large country estate or home. On the other hand he may live in a small terraced house without any front garden at all. In the first case he can intimidate, chase and nip any that encroach upon what he will see as his very large territory, while in the second he may indulge in his behaviour through the window or front door, awaiting the pent up opportunity to display some serious aggression.

Very often it is just nipping, although, combined with high or dominant rank, the severity of the bite often increases. Therefore it is important to ascertain if any other unwanted parts of the dog's character is influencing his behaviour. Fear is a classic combination, with the dog having already learned to move unwanted strangers or animals from his space by going forward aggressively.

An over average predatory instinct can also play a part, for chasing off that which moves is great fun even if there is no aggressive attempt. Therefore a good practitioner will take into account all of these possibilities and treat them at the same time or, if appropriate, before dealing with the territorial behaviour.

Solution

Having ascertained whether the dog in question is acting purely from territorial instinct or whether there are complications of other undesirable behaviours is the first step. Any mix of character that drives a dog to chase will be evident if a thorough investigative consultation has taken place, for other complicating behaviours will have become self evident.

Far be for me to suggest this as a cure, but a change of property can provide a window of opportunity to institute remedial work. When arriving at the new home the old circumstances attached to the habitual behaviour will have been removed.

Territorial dog

Sometimes the desire to change the dog's behaviour coincides with moving as the owners realise their dog's behaviour may give them an immediate bad name with their new neighbours. In the small number of cases where this situation coincides, the opportunity should not be lost of establishing required circumstantial learning in the new home. How this should be done should be learned before moving home.

Psychological: Readers will not be surprised when I refer yet again to the importance of superior rank having been established. While the strength of the dog's driven motivation to be aggressive will need to be taken into account, if the dog understands that he is a pack inferior then the owner is more likely to be able to impress the aura of "if any member of the pack has the right to initiate aggression it is me and not you". Unfortunately, in the majority of cases that will not be the case, and where it might be driven motivation, it will probably far outweigh the dog's deference to superiority. That does not mean that efforts to establish such a comprehension should not be made. Anything that adds another piece to the jigsaw of rehabilitation is important.

Practical: If any dominant behaviour is apparent and the dog's rank is sufficiently high enough to invoke aggression to anyone interfering with his guarding, then practical work certainly will not be an option. If that is the case then clearly there cannot be a safe confrontation with the dog, and the only remedies will be psychological work and avoiding the problem occurring. In some situations, and where the problem is extreme, that may mean that the ultimate option must be advised and a good practitioner should not be afraid to say so. On the other hand it may be that the dog can be kept well away from people, maybe in the owners own kennels, out of the way from possible visitors.

As every situation is different there is no rule of thumb. If the dog is deemed to be one that will back down to a challenge then aversion therapies should be tried to gauge the right one to use. Then it will be a question of how to set everything up so that when the dog is incited to indulge they will be used in the most effective manner.

7

VEHICLE AGGRESSION

In these modern days, so often dogs become involved with cars in one way or another. Many do not have a worrying behaviour associated with vehicles. However, a few do, and ideally the very first signs of a problem should be nipped in the bud. Inevitably such problems do not disappear with time but rather increase in ferocity. One of the problems for a practitioner is that it is often after many months or years of an aggressive habit being constantly reinforced that help is sought.

The problem and causes

There are usually three categories of such behaviour. The first is aggression to those that the dog perceives to be too close to their territory or put part of their body into the vehicle such as a hand or face. That is territorial aggression. Next are the dogs that make mock attacks at vehicles of all types that pass the car they are in. This is usually predatory aggression, but any element of territorial behaviour exacerbating the predatory part of it cannot always be ruled out. Thirdly there is the dog that, when being walked, is aggressive to passing vehicles, and again this is Predatory Aggression.

With vehicle territorial aggression it most usually occurs when the vehicle is stationary with windows or doors partially or fully open. In such situations the dog will attack anyone who passes close enough for him to do so. Even with all windows and doors shut such behaviour can be extremely intimidating and frightening if the passer-by is taken unawares as the dog lunges at the window.

When the problem is the dog travelling by car trying to attack passing vehicles it may not result in anyone being bitten, but the distraction obviously inflicted upon the driver could result in a motor accident.

Solution

Commands alone are unlikely to change this behaviour, nor will distraction be suitable. It would also be dangerous for the driver to be the trainer when the dog's undesirable behaviour is inside a moving vehicle. In such cases an assistant must be the one to be taught by the practitioner.

Psychological: The usual pieces of the rehabilitation jig-saw need to be in place, and that is the dog understanding that he is an inferior member of the pack. Where the dog is also of extreme high rank then this may not be possible

Practical: Aversion therapies are usually the only option but, as in all training, to be effective the chosen therapy must be consistently applied every time the dog indulges. This is again a case of applying it when the first sign of the dog thinking about aggression becomes apparent.

Territorial car aggression

8

TWO BITCHES FIGHTING

Dog to dog aggression can be mild or the opposite. In many cases it will not occur at all for almost always the problem is driven by rank. If the two dogs are of average rank or below then it is unlikely. If one of them is highly ranked and the other sufficiently lower then the latter dog will back down to the higher ranked one and there will not be any fighting. For me such a gap is around a minimum of 20 points, and as that gap becomes progressively less then it is likely that the inferior one may still stake a claim for superiority. That will be as a result of not having perceived that the other dog's mental strength is sufficiently greater.

When there is a situation of close high ranking then fireworks can be expected as neither dog will give way when the other feels they should have done. That is quite a good description of human behaviour but, of course, so much of dog behaviour is reflected by humans with the addition of greater intelligence and usually deviousness added for good measure.

The problem and causes

When two males fight the psychological feeling of superiority will usually be overcome by the physical superiority of the other. Therefore deference is shown by clearly losing the fight and acknowledging it by not continuing to fight back or, perhaps, running away.

Very often such fights can sound extremely ferocious but subsequently, when inspected, other than much saliva there is little evidence of damage to either dog. When that occurs it really is a mild pack order dispute for neither dog is attempting to inflict permanent damage but rather seeks to use his mouth for sound and grip without including the use of full jaw power. In other words it is just a means of settling an argument. The higher ranked one or both of the two dogs are, then the greater will be the damage inflicted by one or both dogs. However, when it is males there usually comes a point when one will consider that the other has shown deference and at that point, or very soon, the fight stops.

With two high ranked bitches fighting that is not the case, and there have been many instances of them fighting to the death. Often this occurs when they are closely confined, such as in a car. Whereas a male's pride can be satisfied with obvious victory, it is not pride that drives females to fight but rather a total dislike of each other. Again, not unlike human behaviour.

Solution

If the problem is two family dogs fighting then whether or not it can be stopped will depend upon the severity of it. Usually the degree of aggression will reflect natural born rank.

Psychological: If one bitch will back down to a much higher ranked bitch then she will also back down to owners who have established ranking superiority, irrespective of their gender. I have dealt with cases where both of the bitches would back down to perceived higher rank and not dare fight when that person was present. However, the problem still remained every time that superior ranked being was removed from the equation. Therefore, while psychological work can be important, it may not continue to work when some circumstances are changed.

Practical: In cases I have dealt with where the bitches have backed down, it has been to noise aversion therapy, but before noise can be used it is necessary to understand any differences in sound sensitivity from one bitch to another. There may well be a difference, and first this must be ascertained.

If success is apparent while the dogs can be supervised, they will need to be securely separated when that is not the case. If it is a case of extreme rank, and aggression cannot immediately be stopped by therapy, then the bitches should never be allowed together again or one must be permanently rehomed to a place where they will not be able to inflict their dominant behaviour on other dogs. For me, keeping one or both bitches muzzled in the house is not an option.

CONCLUSION

In this book I have tried to help those who wish to understand their own or other people's dogs behaviour and also how to deal with it. I hope that I have been successful, and if it helps people and dogs living together to have a better life then I will have achieved my objective.

"All animals are equal but some are more equal than others."
- *George Orwell*

Contentment

Index

The "HOW TO" series by Peter Lewis

HOW TO MAKE GARDEN AGILITY OBSTACLES - Inexpensive and Space Saving
Easy to understand instructions, with illustrations and dimensions, of how to construct inexpensive and space saving agility equipment in your garden.

HOW TO TEACH DOGS AGILITY OBSTACLES - The right way to start Agility
This manual, written by the sport's world renowned innovator of training practices, sets out the best way to start for those who just wish to have fun with the sport or to compete.

HOW TO WIN AT AGILITY
Written by the world's foremost agility expert, this manual teaches such skill for those who want fun or who aspire to be the world champion.

HOW TO JUDGE AGILITY
Peter Lewis is largely credited as having developed the sport and bringing about good judging techniques. This manual explains what every judge needs to know and covers both U.K. and international judging.

HOW TO BECOME A DOG (Understanding dogs)
The aim of this manual is to make owners aware of how a dog's mind works so that they can then act in an appropriate manner that the dog will understand; thus he becomes much easier to train. A must for all pet dog owners.

HOW TO TRAIN YOUR DOG (Part 1 - Household Manners)
This manual and its Part 2 sister are written with the pet dog owner in mind. As a practical dog trainer with years of competition dog training success behind him, the author has spent much of his life teaching people how to avoid problems and successfully train their dog.

HOW TO TRAIN YOUR DOG (Part 2 - Home and Park Training)
This manual recognises that not all training methods are suitable for every dog. It therefore offers in most cases three alternative methods so that the reader can choose one or more which they feel applicable to their dog. It includes exercises such as walking at heel, calling the dog back, sitting, standing and going down when told to do so.

HOW TO HAVE A PROBLEM FREE PUPPY - From Puppy to Well Behaved Dog
Following the advice given in this manual will avoid owners making basic mistakes and result in a problem free puppy.

HOW TO TRAIN A WORKING TRIAL DOG
This 152 page rewrite of Peter's original "The Working Trial Dog" concentrates in detail on foundations and aspects of training from the lowest level to the tracking dog stake. Illustrated with line drawings by the author, it is anticipated that many will read it and take their dog to the top as happened with the original.